LECTURES
ON
MATHEMATICS

LECTURES ON MATHEMATICS

FELIX KLEIN

AMS CHELSEA PUBLISHING
American Mathematical Society • Providence, Rhode Island

2000 *Mathematics Subject Classification.* Primary 01A55, 01A73, 11–XX, 13–XX, 14–XX, 22Exx, 33–XX, 34–XX, 34Mxx, 35–XX, 51M10, 53Axx, 53Dxx.

Library of Congress Cataloging-in-Publication Data

Klein, Felix, 1849–1925.
[Evanston colloquium]
Lectures on mathematics / Felix Klein.
p. cm.
Originally published: The Evanston colloquium. New York : Macmillan, 1894.
Includes bibliographical references.
ISBN 0-8218-2733-2 (alk. paper)
1. Mathematics. I. Title.
QA7.K6 2000
510—dc21 00-058295

Reprinted by the American Mathematical Society, 1911, 2000
∞ The paper used in this book is acid-free and falls within the guidelines established to ensure permanence and durability.
Visit the AMS home page at URL: http://www.ams.org/

10 9 8 7 6 5 4 3 2 1 05 04 03 02 01 00

To reproduce after a lapse of seventeen years lectures which at the time they were delivered were in such close contact with the most recent work of that day, may well call for a word of justification. Has mathematics not advanced since then, and are the questions here treated of first importance at the present time? I reply by asking: What *is* important in the development of mathematics? Is it solely the attainment of new results of potential value, or must not an essential part of the best scientific efforts of each new age be devoted to possessing itself of the heritage of the age that has just preceded it?

It is a noble example that Professor Klein has set all ages of mathematicians that, beginning his activity at a time when the contributions of the immediate past were so rich and so unrelated, he was able to uncover the essential bonds that connect them and to discern the fields to whose development the new methods were best adapted.

His instinct for that which is vital in mathematics is sure, and the light with which his treatment illumines the problems here considered may well serve as a guide for the youth who is approaching the study of the problems of a later day.

WILLIAM F. OSGOOD.

CAMBRIDGE, December 31, 1910.

PREFACE.

The Congress of Mathematics held under the auspices of the World's Fair Auxiliary in Chicago, from the 21st to the 26th of August, 1893, was attended by Professor Felix Klein of the University of Göttingen, as one of the commissioners of the German university exhibit at the Columbian Exposition. After the adjournment of the Congress, Professor Klein kindly consented to hold a *colloquium* on mathematics with such members of the Congress as might wish to participate. The Northwestern University at Evanston, Ill., tendered the use of rooms for this purpose and placed a collection of mathematical books from its library at the disposal of the members of the colloquium. The following is a list of the members attending the colloquium: —

W. W. Beman, A.M., professor of mathematics, University of Michigan.
E. M. Blake, Ph.D., instructor in mathematics, Columbia College.
O. Bolza, Ph.D., associate professor of mathematics, University of Chicago.
H. T. Eddy, Ph.D., president of the Rose Polytechnic Institute.
A. M. Ely, A.B., professor of mathematics, Vassar College.
F. Franklin, Ph.D., professor of mathematics, Johns Hopkins University.
T. F. Holgate, Ph.D., instructor in mathematics, Northwestern University.
L. S. Hulburt, A.M., instructor in mathematics, Johns Hopkins University.
F. H. Loud, A.B., professor of mathematics and astronomy, Colorado College.
J. McMahon, A.M., assistant professor of mathematics, Cornell University.
H. Maschke, Ph.D., assistant professor of mathematics, University of Chicago.
E. H. Moore, Ph.D., professor of mathematics, University of Chicago.

J. E. Oliver, A.M., professor of mathematics, Cornell University.

A. M. Sawin, Sc.M., Evanston.

W. E. Story, Ph.D., professor of mathematics, Clark University.

E. Study, Ph.D., professor of mathematics, University of Marburg.

H. Taber, Ph.D., assistant professor of mathematics, Clark University.

H. W. Tyler, Ph.D., professor of mathematics, Massachusetts Institute of Technology.

J. M. Van Vleck, A.M., LL.D., professor of mathematics and astronomy, Wesleyan University.

E. B. Van Vleck, Ph.D., instructor in mathematics, University of Wisconsin.

C. A. Waldo, A.M., professor of mathematics, De Pauw University.

H. S. White, Ph.D., associate professor of mathematics, Northwestern University.

M. F. Winston, A.B., honorary fellow in mathematics, University of Chicago.

A. Ziwet, assistant professor of mathematics, University of Michigan.

The meetings lasted from August 28th till September 9th; and in the course of these two weeks Professor Klein gave a daily lecture, besides devoting a large portion of his time to personal intercourse and conferences with those attending the meetings. The lectures were delivered freely, in the English language, substantially in the form in which they are here given to the public. The only change made consists in obliterating the conversational form of the frequent questions and discussions by means of which Professor Klein understands so well to enliven his discourse. My notes, after being written out each day, were carefully revised by Professor Klein himself, both in manuscript and in the proofs.

As an appendix it has been thought proper to give a translation of the interesting historical sketch contributed by Professor Klein to the work *Die deutschen Universitäten*. The translation was prepared by Professor H. W. Tyler, of the Massachusetts Institute of Technology.

It is to be hoped that the proceedings of the Chicago Congress of Mathematics, in which Professor Klein took a leading part, will soon be published in full. The papers presented to this Congress, and the discussions that followed their reading, form an important complement to the Evanston colloquium. Indeed, in reading the lectures here published, it should be kept in mind that they followed immediately upon the adjournment of the Chicago meeting, and were addressed to members of the Congress. This circumstance, in addition to the limited time and the informal character of the colloquium, must account for the incompleteness with which the various subjects are treated.

In concluding, the editor wishes to express his thanks to Professors W. W. Beman and H. S. White for aid in preparing the manuscript and correcting the proofs.

ALEXANDER ZIWET.

ANN ARBOR, MICH., November, 1893.

The new edition is printed from the original plates. The only changes in the text consist in the correction of a few misprints.

A. Z.

ANN ARBOR, January, 1911.

CONTENTS.

LECTURES ON MATHEMATICS.

LECTURE I.: CLEBSCH.

(August 28, 1893.)

IT will be the object of our *Colloquia* to pass in review some of the principal phases of the most recent development of mathematical thought in Germany.

A brief sketch of the growth of mathematics in the German universities in the course of the present century has been contributed by me to the work *Die deutschen Universitäten*, compiled and edited by Professor *Lexis* (Berlin, Asher, 1893), for the exhibit of the German universities at the World's Fair.* The strictly objective point of view that had to be adopted for this sketch made it necessary to break off the account about the year 1870. In the present more informal lectures these restrictions both as to time and point of view are abandoned. It is just the period since 1870 that I intend to deal with, and I shall speak of it in a more subjective manner, insisting particularly on those features of the development of mathematics in which I have taken part myself either by personal work or by direct observation.

The first week will be devoted largely to *Geometry*, taking this term in its broadest sense; and in this first lecture it will surely be appropriate to select the celebrated geometer *Clebsch*

---

* A translation of this sketch will be found in the Appendix, p. 99.

as the central figure, partly because he was one of my principal teachers, and also for the reason that his work is so well known in this country.

Among mathematicians in general, three main categories may be distinguished; and perhaps the names *logicians*, *formalists*, and *intuitionists* may serve to characterize them. (1) The word *logician* is here used, of course, without reference to the mathematical logic of Boole, Peirce, etc.; it is only intended to indicate that the main strength of the men belonging to this class lies in their logical and critical power, in their ability to give strict definitions, and to derive rigid deductions therefrom. The great and wholesome influence exerted in Germany by *Weierstrass* in this direction is well known. (2) The *formalists* among the mathematicians excel mainly in the skilful formal treatment of a given question, in devising for it an "algorithm." *Gordan*, or let us say *Cayley* and *Sylvester*, must be ranged in this group. (3) To the *intuitionists*, finally, belong those who lay particular stress on geometrical intuition (*Anschauung*), not in pure geometry only, but in all branches of mathematics. What Benjamin Peirce has called "geometrizing a mathematical question" seems to express the same idea. Lord *Kelvin* and *von Staudt* may be mentioned as types of this category.

Clebsch must be said to belong both to the second and third of these categories, while I should class myself with the third, and also the first. For this reason my account of Clebsch's work will be incomplete; but this will hardly prove a serious drawback, considering that the part of his work characterized by the second of the above categories is already so fully appreciated here in America. In general, it is my intention here, not so much to give a complete account of any subject, as to *supplement* the mathematical views that I find prevalent in this country.

As the first achievement of Clebsch we must set down the introduction into Germany of the work done previously by Cayley and Sylvester in England. But he not only transplanted to German soil their theory of invariants and the interpretation of projective geometry by means of this theory; he also brought this theory into live and fruitful correlation with the fundamental ideas of Riemann's theory of functions. In the former respect, it may be sufficient to refer to Clebsch's *Vorlesungen über Geometrie*, edited and continued by Lindemann; to his *Binäre algebraische Formen*, and in general to what he did in co-operation with Gordan. A good historical account of his work will be found in the biography of Clebsch published in the *Math. Annalen*, Vol. 7.

Riemann's celebrated memoir of 1857* presented the new ideas on the theory of functions in a somewhat startling novel form that prevented their immediate acceptance and recognition. He based the theory of the Abelian integrals and their inverse, the Abelian functions, on the idea of the surface now so well known by his name, and on the corresponding fundamental theorems of existence (*Existenztheoreme*). Clebsch, by taking as his starting-point an algebraic curve defined by its equation, made the theory more accessible to the mathematicians of his time, and added a more concrete interest to it by the geometrical theorems that he deduced from the theory of Abelian functions. Clebsch's paper, *Ueber die Anwendung der Abel'schen Functionen in der Geometrie*,† and the work of Clebsch and Gordan on Abelian functions,‡ are well known to American mathematicians; and in accordance with my plan, I proceed to give merely some critical remarks.

---

* *Theorie der Abel'schen Functionen*, Journal für reine und angewandte Mathematik, Vol. 54 (1857), pp. 115–155; reprinted in Riemann's *Werke*, 1876, pp. 81–135.

† Journal für reine und angewandte Mathematik, Vol. 63 (1864), pp. 189–243.

‡ *Theorie der Abel'schen Functionen*, Leipzig, Teubner, 1866.

However great the achievement of Clebsch's in making the work of Riemann more easy of access to his contemporaries, it is my opinion that at the present time the book of Clebsch is no longer to be considered as the standard work for an introduction to the study of Abelian functions. The chief objections to Clebsch's presentation are twofold: they can be briefly characterized as a lack of mathematical rigour on the one hand, and a loss of intuitiveness, of geometrical perspicuity, on the other. A few examples will explain my meaning.

(*a*) Clebsch bases his whole investigation on the consideration of what he takes to be the most general type of an algebraic curve, and this *general* curve he assumes as having only double points, but no other singularities. To obtain a sure foundation for the theory, it must be proved that any algebraic curve can be transformed rationally into a curve having only double points. This proof was not given by Clebsch; it has since been supplied by his pupils and followers, but the demonstration is long and involved. See the papers by Brill and Nöther in the *Math. Annalen*, Vol. 7 (1874),* and by Nöther, *ib.*, Vol. 23 (1884).†

Another defect of the same kind occurs in connection with the determinant of the periods of the Abelian integrals. This determinant never vanishes as long as the curve is irreducible. But Clebsch and Gordan neglect to prove this; and however simple the proof may be, this must be regarded as an inexactness.

The apparent lack of critical spirit which we find in the work of Clebsch is characteristic of the geometrical epoch in which

* *Ueber die algebraischen Functionen und ihre Anwendung in der Geometrie*, pp. 269–310.

† *Rationale Ausführung der Operationen in der Theorie der algebraischen Functionen*, pp. 311–358.

he lived, the epoch of Steiner, among others. It detracts in nowise from the merit of his work. But the influence of the theory of functions has taught the present generation to be more exacting.

(*b*) The second objection to adopting Clebsch's presentation lies in the fact that, from Riemann's point of view, many points of the theory become far more simple and almost self-evident, whereas in Clebsch's theory they are not brought out in all their beauty. An example of this is presented by the idea of the deficiency p. In Riemann's theory, where p represents the order of connectivity of the surface, the invariability of p under any rational transformation is self-evident, while from the point of view of Clebsch this invariability must be proved by means of a long elimination, without affording the true geometrical insight into its meaning.

For these reasons it seems to me best to begin the theory of Abelian functions with Riemann's ideas, without, however, neglecting to give later the purely algebraical developments. This method is adopted in my paper on Abelian functions;* it is also followed in the work *Die elliptischen Modulfunctionen*, Vols. I. and II., edited by Dr. Fricke. A general account of the historical development of the theory of algebraic curves in connection with Riemann's ideas will be found in my (lithographed) lectures on *Riemann'sche Flächen*, delivered in 1891–92.†

If this arrangement be adopted, it is interesting to follow out the true relation that the algebraical developments bear to Riemann's theory. Thus in Brill and Nöther's theory, the so-called *fundamental theorem* of Nöther is of primary impor-

---

* *Zur Theorie der Abel'schen Functionen*, Math. Annalen, Vol. 36 (1890), pp. 1–83.

† My lithographed lectures frequently give only an outline of the subject, omitting details and long demonstrations, which are supposed to be supplied by the student by private reading and a study of the literature of the subject.

tance. It gives a rule for deciding under what conditions an algebraic rational integral function f of x and y can be put into the form

$$f = A\phi + B\psi,$$

where ϕ and ψ are likewise rational algebraic functions. Each point of intersection of the curves $\phi = 0$ and $\psi = 0$ must of course be a point of the curve $f = 0$. But there remains the question of multiple and singular points; and this is disposed of by Nöther's theorem. Now it is of great interest to investigate how these relations present themselves when the starting-point is taken from Riemann's ideas.

One of the best illustrations of the utility of adopting Riemann's principles is presented by the very remarkable advance made recently by Hurwitz, in the theory of algebraic curves, in particular his extension of the theory of algebraic correspondences, an account of which is given in the second volume of the *Elliptische Modulfunctionen*. Cayley had found as a fundamental theorem in this theory a rule for determining the number of self-corresponding points for algebraic correspondences of a simple kind. A whole series of very valuable papers by Brill, published in the *Math. Annalen*,* is devoted to the further investigation and demonstration of this theorem. Now Hurwitz, attacking the problem from the point of view of Riemann's ideas, arrives not only at a more simple and quite general demonstration of Cayley's rule, but proceeds to a complete study of all possible algebraic correspondences. He finds that while for *general* curves the correspondences consid-

---

* *Ueber zwei Berührungsprobleme*, Vol. 4 (1871), pp. 527–549. — *Ueber Entsprechen von Punktsystemen auf einer Curve*, Vol. 6 (1873), pp. 33–65. — *Ueber die Correspondenzformel*, Vol. 7 (1874), pp. 607–622. — *Ueber algebraische Correspondenzen*, Vol. 31 (1888), pp. 374–409. — *Ueber algebraische Correspondenzen. Zweite Abhandlung: Specialgruppen von Punkten einer algebraischen Curve*, Vol. 36 (1890), pp. 321–360.

ered by Cayley and Brill are the only ones that exist, in the case of *singular* curves there are other correspondences which also can be treated completely. These singular curves are characterized by certain linear relations with integral coefficients, connecting the periods of their Abelian integrals.

Let us now turn to that side of Clebsch's method which appears to me to be the most important, and which certainly must be recognized as being of great and permanent value; I mean the generalization, obtained by Clebsch, of the whole theory of Abelian integrals to a theory of algebraic functions with several variables. By applying the methods he had developed for functions of the form $f(x, y) = 0$, or in homogeneous co-ordinates, $f(x_1, x_2, x_3) = 0$, to functions with four homogeneous variables $f(x_1, x_2, x_3, x_4) = 0$, he found in 1868, that there also exists a number p that remains invariant under all rational transformations of the surface $f=0$. Clebsch arrives at this result by considering *double integrals* belonging to the surface.

It is evident that this theory could not have been found from Riemann's point of view. There is no difficulty in conceiving a four-dimensional Riemann space corresponding to an equation $f(x, y, z)=0$. But the difficulty would lie in proving the "theorems of existence" for such a space; and it may even be doubted whether analogous theorems hold in such a space.

While to Clebsch is due the fundamental idea of this grand generalization, the working out of this theory was left to his pupils and followers. The work was mainly carried on by Nöther, who showed, in the case of algebraic surfaces, the existence of more than one invariant number p and of corresponding moduli, *i.e.* constants not changed by one-to-one transformations. Italian and French mathematicians, in particular Picard and Poincaré, have also contributed largely to the further development of the theory.

If the value of a man of science is to be gauged not by his general activity in all directions, but solely by the fruitful new ideas that he has first introduced into his science, then the theory just considered must be regarded as the most valuable work of Clebsch.

In close connection with the preceding are the general ideas put forth by Clebsch in his last memoir,* ideas to which he himself attached great importance. This memoir implies an application, as it were, of the theory of Abelian functions to the theory of differential equations. It is well known that the central problem of the whole of modern mathematics is the study of the transcendental functions defined by differential equations. Now Clebsch, led by the analogy of his theory of Abelian integrals, proceeds somewhat as follows. Let us consider, for example, an ordinary differential equation of the first order $f(x, y, y')=0$, where f represents an algebraic function. Regarding y' as a third variable z, we have the equation of an algebraic surface. Just as the Abelian integrals can be classified according to the properties of the fundamental curve that remain unchanged under a rational transformation, so Clebsch proposes to classify the transcendental functions defined by the differential equations according to the invariant properties of the corresponding surfaces $f=0$ under rational one-to-one transformations.

The theory of differential equations is just now being cultivated very extensively by French mathematicians; and some of them proceed precisely from this point of view first adopted by Clebsch.

---

* *Ueber ein neues Grundgebilde der analytischen Geometrie der Ebene*, Math. Annalen, Vol. 6 (1873), pp. 203-215.

LECTURE II.: SOPHUS LIE.

(August 29, 1893.)

To fully understand the mathematical genius of Sophus Lie, one must not turn to the books recently published by him in collaboration with Dr. Engel, but to his earlier memoirs, written during the first years of his scientific career. There Lie shows himself the true geometer that he is, while in his later publications, finding that he was but imperfectly understood by the mathematicians accustomed to the analytical point of view, he adopted a very general analytical form of treatment that is not always easy to follow.

Fortunately, I had the advantage of becoming intimately acquainted with Lie's ideas at a very early period, when they were still, as the chemists say, in the "nascent state," and thus most effective in producing a strong reaction. My lecture to-day will therefore be devoted chiefly to his paper, "*Ueber Complexe, insbesondere Linien- und Kugel-Complexe, mit Anwendung auf die Theorie partieller Differentialgleichungen.*" *

To define the place of this paper in the historical development of geometry, a word must be said of two eminent geometers of an earlier period: Plücker (1801–68) and Monge (1746–1818). Plücker's name is familiar to every mathematician, through his formulæ relating to algebraic curves. But what is of importance in the present connection is his generalized idea

* *Math. Annalen*, Vol. 5 (1872), pp. 145–256.

of the space-element. The ordinary geometry with the point as element deals with space as three-dimensioned, conformably to the three constants determining the position of a point. A dual transformation gives the plane as element; space in this case has also three dimensions, as there are three independent constants in the equation of the plane. If, however, the straight line be selected as space-element, space must be considered as four-dimensional, since four independent constants determine a straight line. Again, if a quadric surface F_2 be taken as element, space will have nine dimensions, because every such element requires nine quantities for its determination, viz. the nine independent constants of the surface F_2; in other words, space contains ∞^9 quadric surfaces. This conception of hyper-spaces must be clearly distinguished from that of Grassmann and others. Plücker, indeed, rejected any other idea of a space of more than three dimensions as too abstruse. — The work of Monge that is here of importance, is his *Application de l'analyse à la géométrie*, 1809 (reprinted 1850), in which he treats of ordinary and partial differential equations of the first and second order, and applies these to geometrical questions such as the curvature of surfaces, their lines of curvature, geodesic lines, etc. The treatment of geometrical problems by means of the differential and integral calculus is one feature of this work; the other, perhaps even more important, is the converse of this, viz. the application of geometrical intuition to questions of analysis.

Now this last feature is one of the most prominent characteristics of Lie's work; he increases its power by adopting Plücker's idea of a generalized space-element and extending this fundamental conception. A few examples will best serve to give an idea of the character of his work; as such an example I select (as I have done elsewhere before) Lie's sphere-geometry (*Kugelgeometrie*).

Taking the equation of a sphere in the form

$$x^2+y^2+z^2-2Bx-2Cy-2Dz+E=0,$$

the coefficients, B, C, D, E, can be regarded as the co-ordinates of the sphere, and ordinary space appears accordingly as a manifoldness of four dimensions. For the radius, R, of the sphere we have

$$R^2=B^2+C^2+D^2-E$$

as a relation connecting the fifth quantity, R, with the four co-ordinates, B, C, D, E.

To introduce homogeneous co-ordinates, put

$$B=\frac{b}{a},\quad C=\frac{c}{a},\quad D=\frac{d}{a},\quad E=\frac{e}{a},\quad R=\frac{r}{a};$$

then $a:b:c:d:e$ are the five homogeneous co-ordinates of the sphere, and the sixth quantity r is related to them by means of the homogeneous equation of the second degree,

$$r^2=b^2+c^2+d^2-ae. \qquad (1)$$

Sphere-geometry has been treated in two ways that must be carefully distinguished. In one method, which we may call *the elementary sphere-geometry*, only the five co-ordinates $a:b:c:d:e$ are used, while in the other, *the higher*, or *Lie's*, *sphere-geometry*, the quantity r is introduced. In this latter system, a sphere has six homogeneous co-ordinates, a, b, c, d, e, r, connected by the equation (1).

From a higher point of view the distinction between these two sphere-geometries, as well as their individual character, is best brought out by considering the *group* belonging to each. Indeed, every system of geometry is characterized by its group, in the meaning explained in my Erlangen *Programm*;* *i.e.*

---

* *Vergleichende Betrachtungen über neuere geometrische Forschungen. Programm zum Eintritt in die philosophische Facultät und den Senat der K. Friedrich-Alexan-*

every system of geometry deals only with such relations of space as remain unchanged by the transformations of its group.

In the elementary sphere-geometry the group is formed by all the linear substitutions of the five quantities a, b, c, d, e, that leave unchanged the homogeneous equation of the second degree

$$b^2 + c^2 + d^2 - ae = 0. \tag{2}$$

This gives $\infty^{25-15} = \infty^{10}$ substitutions. By adopting this definition we obtain point-transformations of a simple character. The geometrical meaning of equation (2) is that the radius is zero. Every sphere of vanishing radius, *i.e.* every point, is therefore transformed into a point. Moreover, as the polar

$$2\,bb' + 2\,cc' + 2\,dd' - ae' - a'e = 0$$

remains likewise unchanged in the transformation, it follows that orthogonal spheres are transformed into orthogonal spheres. Thus the group of the elementary sphere-geometry is characterized as the *conformal group*, well known as that of the transformation by inversion (or reciprocal radii) and through its applications in mathematical physics.

Darboux has further developed this elementary sphere-geometry. Any equation of the second degree

$$F(a, b, c, d, e) = 0,$$

taken in connection with the relation (2) represents a point-surface which Darboux has called *cyclide*. From the point of view of ordinary projective geometry, the cyclide is a surface of the fourth order containing the imaginary circle common to all spheres of space as a double curve. A careful investigation

---

ders-Universität zu Erlangen. Erlangen, Deichert, 1872. For an English translation, by Haskell, see the Bulletin of the New York Mathematical Society, Vol. 2 (1893), pp. 215–249.

of these cyclides will be found in Darboux's *Leçons sur la théorie générale des surfaces et les applications géométriques du calcul infinitésimal*, and elsewhere. As the ordinary surfaces of the second degree can be regarded as special cases of cyclides, we have here a method for generalizing the known properties of quadric surfaces by extending them to cyclides. Thus Mr. M. Bôcher, of Harvard University, in his dissertation,* has treated the extension of a problem in the theory of the potential from the known case of a body bounded by surfaces of the second degree to a body bounded by cyclides. A more extended publication on this subject by Mr. Bôcher will appear in a few months (Leipzig, Teubner).

In the higher sphere-geometry of Lie, the six homogeneous co-ordinates $a:b:c:d:e:r$ are connected, as mentioned above, by the homogeneous equation of the second degree,

$$b^2+c^2+d^2-r^2-ae=0.$$

The corresponding group is selected as the group of the linear substitutions transforming this equation into itself. We have thus a group of $\infty^{36-21}=\infty^{15}$ substitutions. But this is not a group of point-transformations; for a sphere of radius zero becomes a sphere whose radius is in general different from zero. Thus, putting for instance

$$B'=B,\ C'=C,\ D'=D,\ E'=E,\ R'=R+\text{const.},$$

it appears that the transformation consists in a mere dilatation or expansion of each sphere, a point becoming a sphere of given radius.

The meaning of the polar equation

$$2\,bb'+2\,cc'+2\,dd'-2\,rr'-ae'-a'e=0$$

---

* *Ueber die Reihenentwickelungen der Potentialtheorie*, gekrönte Preisschrift, Göttingen, Dieterich, 1891.

remaining invariant for any transformation of the group, is evidently that the spheres originally in contact remain in contact. The group belongs therefore to the important class of *contact-transformations*, which will soon be considered more in detail.

In studying any particular geometry, such as Lie's sphere-geometry, two methods present themselves.

(1) We may consider equations of various degrees and inquire what they represent. In devising names for the different configurations so obtained, Lie used the names introduced by Plücker in his line-geometry. Thus a single equation,

$$F(a, b, c, d, e, r) = 0,$$

is said to represent a *complex* of the first, second, etc., degree, according to the degree of the equation; a complex contains, therefore, ∞^3 spheres. Two such equations,

$$F_1 = 0, \qquad F_2 = 0,$$

represent a *congruency* containing ∞^2 spheres. Three equations,

$$F_1 = 0, \quad F_2 = 0, \quad F_3 = 0,$$

may be said to represent a *set* of spheres, the number being ∞^1. It is to be noticed that in each case the equation of the second degree,

$$b^2 + c^2 + d^2 - r^2 - ae = 0,$$

is understood to be combined with the equation $F = 0$.

It may be well to mention expressly that the same names are used by other authors in the elementary sphere-geometry, where their meaning is, of course, different.

(2) The other method of studying a new geometry consists in inquiring how the ordinary configurations of point-geometry can be treated by means of the new system. This line of inquiry has led Lie to highly interesting results.

In ordinary geometry a surface is conceived as a locus of points; in Lie's geometry it appears as the totality of all the spheres having contact with the surface. This gives a threefold infinity of spheres, or a complex of spheres,

$$F(a, b, c, d, e, r) = 0.$$

But this, of course, is not a *general* complex; for not every complex will be such as to touch a surface. It has been shown that the condition that must be fulfilled by a complex of spheres, if all its spheres are to touch a surface, is the following:

$$\left(\frac{\partial F}{\partial b}\right)^2 + \left(\frac{\partial F}{\partial c}\right)^2 + \left(\frac{\partial F}{\partial d}\right)^2 - \left(\frac{\partial F}{\partial r}\right)^2 - \frac{\partial F}{\partial a}\frac{\partial F}{\partial e} = 0.$$

To give at least one illustration of the further development of this interesting theory, I will mention that among the infinite number of spheres touching the surface at any point there are two having stationary contact with the surface; they are called the *principal spheres.* The lines of curvature of the surface can then be defined as curves along which the principal spheres touch the surface in two successive points.

Plücker's line-geometry can be studied by the same two methods just mentioned. In this geometry let $p_{12}, p_{13}, p_{14}, p_{34}, p_{42}, p_{23}$ be the usual six homogeneous co-ordinates, where $p_{ik} = -p_{ki}$. Then we have the identity

$$p_{12}p_{34} + p_{13}p_{42} + p_{14}p_{23} = 0,$$

and we take as group the ∞^{15} linear substitutions transforming this equation into itself. This group corresponds to the totality of collineations and reciprocations, *i.e.* to the projective group. The reason for this lies in the fact that the polar equation

$$p_{12}p_{34}' + p_{13}p_{42}' + p_{14}p_{23}' + p_{34}p_{12}' + p_{42}p_{13}' + p_{23}p_{14}' = 0$$

expresses the intersection of the two lines p, p'.

Now Lie has instituted a comparison of the highest interest between the line-geometry of Plücker and his own sphere-geometry. In each of these geometries there occur six homogeneous co-ordinates connected by a homogeneous equation of the second degree. The discriminant of each equation is different from zero. It follows that we can pass from either of these geometries to the other by linear substitutions. Thus, to transform

$$p_{12}p_{34}+p_{13}p_{42}+p_{14}p_{23}=0$$

into

$$b^2+c^2+d^2-r^2-ae=0,$$

it is sufficient to assume, say,

$$p_{12}=b+ic, \quad p_{13}=d+r, \quad p_{14}=-a,$$

$$p_{34}=b-ic, \quad p_{42}=d-r, \quad p_{23}=e.$$

It follows from the linear character of the substitutions that the polar equations are likewise transformed into each other. Thus we have the remarkable result that *two spheres that touch correspond to two lines that intersect.*

It is worthy of notice that the equations of transformation involve the imaginary unit i; and the law of inertia of quadratic forms shows at once that this introduction of the imaginary cannot be avoided, but is essential.

To illustrate the value of this transformation of line-geometry into sphere-geometry, and *vice versa,* let us consider three linear equations,

$$F_1=0, \ F_2=0, \ F_3=0,$$

the variables being either line co-ordinates or sphere co-ordinates. In the former case the three equations represent a *set of lines; i.e.* one of the two sets of straight lines of a hyperboloid of one sheet. It is well known that each line of either set intersects all the lines of the other. Transforming to sphere-geometry, we obtain a *set of spheres* corresponding to each set of lines; and every sphere of either set must touch every sphere of the other set. This gives a configuration well known in geometry from other investigations; viz. all these spheres envelop a surface known as Dupin's cyclide. We have thus found a noteworthy correlation between the hyperboloid of one sheet and Dupin's cyclide.

Perhaps the most striking example of the fruitfulness of this work of Lie's is his discovery that by means of this transformation the lines of curvature of a surface are transformed into asymptotic lines of the transformed surface, and *vice versa.* This appears by taking the definition given above for the lines of curvature and translating it word for word into the language of line-geometry. Two problems in the infinitesimal geometry of surfaces, that had long been regarded as entirely distinct, are thus shown to be really identical. This must certainly be regarded as one of the most elegant contributions to differential geometry made in recent times.

LECTURE III.: SOPHUS LIE.

(August 30, 1893.)

THE distinction between analytic and algebraic functions, so important in pure analysis, enters also into the treatment of geometry.

Analytic functions are those that can be represented by a power series, convergent within a certain region bounded by the so-called circle of convergence. Outside of this region the analytic function is not regarded as given *a priori;* its continuation into wider regions remains a matter of special investigation and may give very different results, according to the particular case considered.

On the other hand, an *algebraic* function, $w = \text{Alg.}\ (z)$, is supposed to be known for the whole complex plane, having a finite number of values for every value of z.

Similarly, in geometry, we may confine our attention to a limited portion of an analytic curve or surface, as, for instance, in constructing the tangent, investigating the curvature, etc.; or we may have to consider the whole extent of algebraic curves and surfaces in space.

Almost the whole of the applications of the differential and integral calculus to geometry belongs to the former branch of geometry; and as this is what we are mainly concerned with in the present lecture, we need not restrict ourselves to algebraic functions, but may use the more general analytic functions confining ourselves always to limited portions of space. I thought it advisable to state this here once for all, since here in America the consideration of algebraic curves has perhaps been too predominant.

The possibility of introducing new elements of space has been pointed out in the preceding lecture. To-day we shall use again a new space-element, consisting of an infinitesimal portion of a surface (or rather of its tangent plane) with a definite point in it. This is called, though not very properly, a *surface-element* (*Flächenelement*), and may perhaps be likened to an infinitesimal fish-scale. From a more abstract point of view it may be defined as simply the combination of a plane with a point in it.

As the equation of a plane passing through a point (x, y, z) can be written in the form

$$z' - z = p(x' - x) + q(y' - y),$$

x', y', z' being the current co-ordinates, we have x, y, z, p, q as the co-ordinates of our surface-element, so that space becomes a fivefold manifoldness. If homogeneous co-ordinates be used, the point (x_1, x_2, x_3, x_4) and the plane (u_1, u_2, u_3, u_4) passing through it are connected by the condition

$$x_1 u_1 + x_2 u_2 + x_3 u_3 + x_4 u_4 = 0,$$

expressing their united position; and the number of independent constants is $3+3-1=5$, as before.

Let us now see how ordinary geometry appears in this representation. A point, being the locus of all surface-elements passing through it, is represented as a manifoldness of two dimensions, let us say for shortness, an M_2. A curve is represented by the totality of all those surface-elements that have their point on the curve and their plane passing through the tangent; these elements form again an M_2. Finally, a surface is given by those surface-elements that have their point on the surface and their plane coincident with the tangent plane of the surface; they, too, form an M_2.

Moreover, all these M_2's have an important property in common: any two consecutive surface-elements belonging to the same point, curve, or surface always satisfy the condition

$$dz - pdx - qdy = 0,$$

which is a simple case of a Pfaffian relation; and conversely, if two surface-elements satisfy this condition, they belong to the same point, curve, or surface, as the case may be.

Thus we have the highly interesting result that in the geometry of surface-elements points as well as curves and surfaces are brought under one head, being all represented by twofold manifoldnesses having the property just explained. This definition is the more important as there are no other M_2's having the same property.

We now proceed to consider the very general kind of transformations called by Lie *contact-transformations*. They are transformations that change our element (x, y, z, p, q) into (x', y', z', p', q') by such substitutions

$$x' = \phi(x, y, z, p, q), \quad y' = \psi(x, y, z, p, q), \quad z' = \cdots, \quad p' = \cdots, \quad q' = \cdots,$$

as will transform into itself the linear differential equation

$$dz - pdx - qdy = 0.$$

The geometrical meaning of the transformation is evidently that any M_2 having the given property is changed into an M_2 having the same property. Thus, for instance, a surface is transformed generally into a surface, or in special cases into a point or a curve. Moreover, let us consider two manifoldnesses M_2 having a contact, *i.e.* having a surface-element in common; these M_2's are changed by the transformation into two other M_2's having also a contact. From this characteristic the name given by Lie to the transformation will be understood.

Contact-transformations are so important, and occur so frequently, that particular cases attracted the attention of geometers long ago, though not under this name and from this point of view, *i.e.* not as contact-transformations, so that the true insight into their nature could not be obtained.

Numerous examples of contact-transformations are given in my (lithographed) lectures on *Höhere Geometrie,* delivered during the winter-semester of 1892–93. Thus, an example in two dimensions is found in the problem of wheel-gearing. The outline of the tooth of one wheel being given, it is here required to find the outline of the tooth of the other wheel, as I explained to you in my lecture at the Chicago Exhibition, with the aid of the models in the German university exhibit.

Another example is found in the theory of perturbations in astronomy; Lagrange's method of variation of parameters as applied to the problem of three bodies is equivalent to a contact-transformation in a higher space.

The group of ∞^{15} substitutions considered yesterday in line-geometry is also a group of contact-transformations, both the collineations and reciprocations having this character. The reciprocations give the first well-known instance of the transformation of a point into a plane (*i.e.* a surface), and a curve into a developable (*i.e.* also a surface). These transformations of curves will here be considered as transforming the *elements* of the points or curves into the *elements* of the surface.

Finally, we have examples of contact-transformations, not only in the transformations of spheres discussed in the last lecture, but even in the general transition from the line-geometry of Plücker to the sphere-geometry of Lie. Let us consider this last case somewhat more in detail.

First of all, two lines that intersect have, of course, a surface-element in common; and as the two corresponding spheres must also have a surface-element in common, they will be in contact, as is actually the case for our transformation. It will be of interest to consider more closely the correlation between the surface-elements of a line and those of a sphere, although it is given by imaginary formulæ. Take, for instance, the totality of the surface-elements belonging to a circle on one of the spheres; we may call this a *circular set* of elements. In line-geometry there corresponds the set of surface-elements along a generating line of a skew surface; and so on. The theorem regarding the transformation of the curves of curvature into asymptotic lines becomes now self-evident. Instead of the curve of curvature of a surface we have here to consider the corresponding elements of the surface which we may call a *curvature set.* Similarly, an asymptotic line is replaced by the elements of the surface along this line; to this the name *osculating set* may be given. The correspondence between the two sets is brought out immediately by considering that two consecutive elements of a curvature set belong to the same sphere, while two consecutive elements of an osculating set belong to the same straight line.

One of the most important applications of contact-transformations is found in the theory of partial differential equations; I shall here confine myself to partial differential equations of the first order. From our new point of view, this theory assumes a much higher degree of perspicuity, and the true meaning of the terms "solution," "general solution," "complete solution," "singular solution," introduced by Lagrange and Monge, is brought out with much greater clearness.

Let us consider the partial differential equation of the first order

$$f(x, y, z, p, q) = 0.$$

In the older theory, a distinction is made according to the way in which p and q enter into the equation. Thus, when p and q enter only in the first degree, the equation is called linear. If p and q should happen to be both absent, the equation would not be regarded as a differential equation at all. From the higher point of view of Lie's new geometry, this distinction disappears entirely, as will be seen in what follows.

The number of all surface elements in the whole of space is of course ∞^5. By writing down our equation we single out from these a manifoldness of four dimensions, M_4, of ∞^4 elements. Now, to find a "solution" of the equation in Lie's sense means to single out from this M_4 a twofold manifoldness, M_2, of the characteristic property; whether this M_2 be a point, a curve, or a surface, is here regarded as indifferent. What Lagrange calls finding a "complete solution" consists in dividing the M_4 into ∞^2 M_2's. This can of course be done in an infinite number of ways. Finally, if any singly infinite set be taken out of the ∞^2 M_2's, we have in the envelope of this set what Lagrange calls a "general solution." These formulations hold quite generally for *all* partial differential equations of the first order, even for the most specialized forms.

To illustrate, by an example, in what sense an equation of the form $f(x, y, z)=0$ may be regarded as a partial differential equation and what is the meaning of its solutions, let us consider the very special case $z=0$. While in ordinary co-ordinates this equation represents all the *points* of the xy-plane, in Lie's system it represents of course all the *surface-elements* whose points lie in the plane. Nothing is so simple as to assign a "complete solution" in this case; we have only to take the ∞^2 points of the plane themselves, each point being an M_2 of the equation. To derive from this the "general solution," we must take all possible singly infinite sets of points in the plane, *i.e.* any curve whatever, and form the envelope of the surface-elements belonging to the points ; in other words, we must take the elements touching the curve. Finally, the plane itself represents of course a "singular solution."

Now, the very high interest and importance of this simple illustration lies in the fact that by a contact-transformation every partial differential equation of the first order can be changed into this particular form $z=0$. Hence the whole disposition of the solutions outlined above holds quite generally.

A new and deeper insight is thus gained through Lie's theory into the meaning of problems that have long been regarded as classical, while at the same time a full array of new problems is brought to light and finds here its answer.

It can here only be briefly mentioned that Lie has done much in applying similar principles to the theory of partial differential equations of the second order.

At the present time Lie is best known through his theory of continuous groups of transformations, and at first glance it might appear as if there were but little connection between this theory and the geometrical considerations that engaged our attention in the last two lectures. I think it therefore desirable to point out here this connection. *It has been the final aim of Lie from the beginning to make progress in the theory of differential equations;* and as subsidiary to this end may be regarded both the geometrical developments considered in these lectures and the theory of continuous groups.

For further particulars concerning the subjects of the present as well as the two preceding lectures, I may refer to my (lithographed) lectures on *Höhere Geometrie,* delivered at Göttingen, in 1892–93. The theory of surface-elements is also fully developed in the second volume of the *Theorie der Transformationsgruppen,* by Lie and Engel (Leipzig, Teubner, 1890).

LECTURE IV.: ON THE REAL SHAPE OF ALGEBRAIC CURVES AND SURFACES.

(August 31, 1893.)

WE turn now to *algebraic* functions, and in particular to the question of the actual geometric forms corresponding to such functions. The question as to the reality of geometric forms and the actual shape of algebraic curves and surfaces was somewhat neglected for a long time. Otherwise it would be difficult to explain, for instance, why the connection between Cayley's theory of projective measurement and the non-Euclidean geometry should not have been perceived at once. As these questions are even now less well known than they deserve to be, I proceed to give here an historical sketch of the subject, without, however, attempting completeness.

It must be counted among the lasting merits of Sir Isaac Newton that he first investigated the shape of the plane curves of the third order. His *Enumeratio linearum tertii ordinis* * shows that he had a very clear conception of projective geometry; for he says that all curves of the third order can be derived by central projection from five fundamental types (Fig. 1). But I wish to direct your particular attention to the paper by Möbius, *Ueber die Grundformen der Linien der dritten Ordnung*,† where the forms of the cubic curves are derived by

* First published as an appendix to Newton's *Opticks*, 1704.

† Abhandlungen der Königl. Sächsischen Gesellschaft der Wissenschaften, math.-phys. Klasse, Vol. I (1852), pp. 1–82; reprinted in Möbius' *Gesammelte Werke*, Vol. II (1886), pp. 89–176.

purely geometric considerations. Owing to its remarkable elegance of treatment, this paper has given the impulse to all the subsequent researches in this line that I shall have to mention.

In 1872 we considered, in Göttingen, the question as to the shape of surfaces of the third order. As a particular case, Clebsch at this time constructed his beautiful model of the

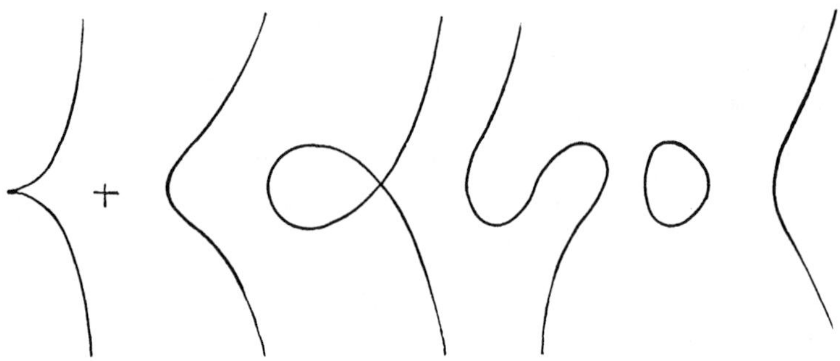

Fig. 1.

diagonal surface, with 27 real lines, which I showed to you at the Exhibition. The equation of this surface may be written in the simple form

$$\sum_1^5 x_i = 0, \quad \sum_1^5 x_i^3 = 0,$$

which shows that the surface can be transformed into itself by the 120 permutations of the x's.

It may here be mentioned as a general rule, that in selecting a particular case for constructing a model the first prerequisite is regularity. By selecting a symmetrical form for the model, not only is the execution simplified, but what is of more importance, the model will be of such a character as to impress itself readily on the mind.

Instigated by this investigation of Clebsch, I turned to the general problem of determining all possible forms of cubic surfaces.* I established the fact that by the principle of continuity all forms of real surfaces of the third order can be derived from the particular surface having four real conical points. This surface, also, I exhibited to you at the World's Fair, and pointed out how the diagonal surface can be derived from it. But what is of primary importance is the completeness of enumeration resulting from my point of view; it would be of comparatively little value to derive any number of special forms if it cannot be proved that the method used exhausts the subject. Models of the typical cases of all the principal forms of cubic surfaces have since been constructed by Rodenberg for Brill's collection.

In the 7th volume of the *Math. Annalen* (1874) Zeuthen† has discussed the various forms of plane curves of the fourth order (C_4). He considers in particular the reality of the double tangents on these curves. The number of such tangents is 28, and they are all real when the curve consists of four separate closed portions (Fig. 2). What is of particular interest is the relation of Zeuthen's researches on quartic curves to my own researches on cubic surfaces, as explained by Zeuthen himself.‡ It had been observed before, by Geiser, that if a cubic surface be projected on a plane from a point on the surface, the contour of the projection is a quartic curve, and that every quartic curve can be generated in this way. If a surface with four conical points be chosen, the resulting quartic has four double points; that is, it breaks up into two conics

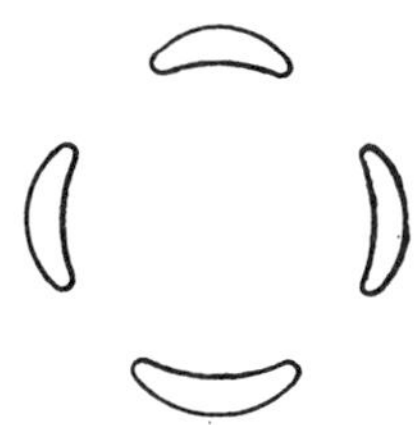

Fig. 2.

* See my paper *Ueber Flächen dritter Ordnung*, Math. Annalen, Vol. 6 (1873), pp. 551–581.

† *Sur les différentes formes des courbes planes du quatrième ordre*, pp. 410–432.

‡ *Études des propriétés de situation des surfaces cubiques*, Math. Annalen, Vol. 8 (1875), pp. 1–30.

(Fig. 3). By considering the shaded portions in the figure it will readily be seen how, by the principle of continuity, the four ovals of the quartic (Fig. 2) are obtained. This corresponds exactly to the derivation of the diagonal surface from the cubic surface having four conical points.

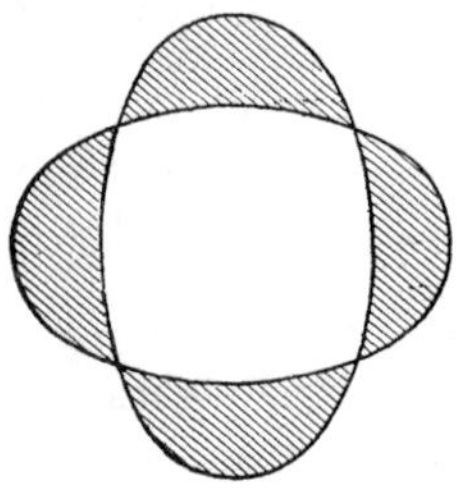

Fig. 3.

The attempts to extend this application of the principle of continuity so as to gain an insight into the shape of curves of the nth order have hitherto proved futile, as far as a general classification and an enumeration of all fundamental forms is concerned. Still, some important results have been obtained. A paper by Harnack* and a more recent one by Hilbert† are here to be mentioned. Harnack finds that, if p be the deficiency of the curve, the maximum number of separate branches the curve can have is $p+1$; and a curve with $p+1$ branches actually exists. Hilbert's paper contains a large number of interesting special results which from their nature cannot be included in the present brief summary.

I myself have found a curious relation between the numbers of real singularities.‡ Denoting the order of the curve by n, the class by k, and considering only simple singularities, we may have three kinds of double points, say d' ordinary and d'' isolated real double points, besides imaginary double points; then there may be r' real cusps, besides imaginary cusps; and similarly, by the principle of duality, t' ordinary, t'' isolated

---

* *Ueber die Vieltheiligkeit der ebenen algebraischen Curven*, Math. Annalen, Vol. 10 (1876), pp. 189–198.

† *Ueber die reellen Züge algebraischer Curven*, Math. Annalen, Vol. 38 (1891), pp. 115–138.

‡ *Eine neue Relation zwischen den Singularitäten einer algebraischen Curve*, Math. Annalen, Vol. 10 (1876), pp. 199–209.

real double tangents, besides imaginary double tangents; also w' real inflexions, besides imaginary inflexions. Then it can be proved by means of the principle of continuity, that the following relation must hold:

$$n + w' + 2t'' = k + r' + 2d''.$$

This general law contains everything that is known as to curves of the third or fourth orders. It has been somewhat extended in a more algebraic sense by several writers. Moreover, Brill, in Vol. 16 of the *Math. Annalen* (1880),* has shown how the formula must be modified when higher singularities are involved.

As regards quartic surfaces, Rohn has investigated an enormous number of special cases; but a complete enumeration he has not reached. Among the special surfaces of the fourth order the Kummer surface with 16 conical points is one of the most important. The models constructed by Plücker in connection with his theory of complexes of lines all represent special cases of the Kummer surface. Some types of this surface are also included in the Brill collection. But all these models are now of less importance, since Rohn found the following interesting and comprehensive result.

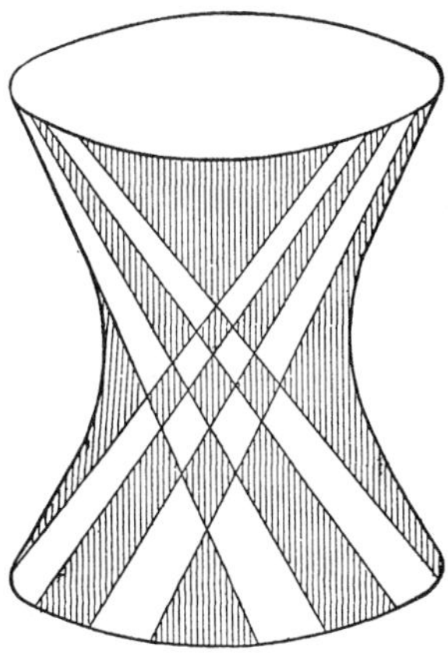

Fig. 4.

Imagine a quadric surface with four generating lines of each set (Fig. 4). According to the character of the surface and the reality, non-reality, or coincidence of these lines, a large number of special cases is possible; all these cases, however, must be

* *Ueber Singularitäten ebener algebraischer Curven und eine neue Curvenspecies*, pp. 348–408.

treated alike. We may here confine ourselves to the case of an hyperboloid of one sheet with four distinct lines of each set. These lines divide the surface into 16 regions. Shading the alternate regions as in the figure, and regarding the shaded regions as double, the unshaded regions being disregarded, we have a surface consisting of eight separate closed portions hanging together only at the points of intersection of the lines; and this is a Kummer surface with 16 real double points. Rohn's researches on the Kummer surface will be found in the *Math. Annalen*, Vol. 18 (1881);* his more general investigations on quartic surfaces, *ib.*, Vol. 29 (1887).†

There is still another mode of dealing with the shape of curves (not of surfaces), viz. by means of the theory of Riemann. The first problem that here presents itself is to establish the connection between a plane curve and a Riemann surface, as I have done in Vol. 7 of the *Math. Annalen* (1874).‡ Let us consider a cubic curve; its deficiency is $p=1$. Now it is well known that in Riemann's theory this deficiency is a measure of the connectivity of the corresponding Riemann surface, which, therefore, in the present case, must be that of a *torus*, or anchor-ring. The question then arises: what has the anchor-ring to do with the cubic curve? The connection will best be understood by considering the curve of the third *class* whose shape is represented in Fig. 5. It is easy to see that of the three tangents that can be drawn to this curve from any point in its plane, all three will be real if the point be selected outside the oval branch, or inside the triangular branch; but that only one of the three tangents will be real for any point in the shaded region, while the other two tangents are imaginary. As

---

* *Die verschiedenen Gestalten der Kummer'schen Fläche*, pp. 99–159.

† *Die Flächen vierter Ordnung hinsichtlich ihrer Knotenpunkte und ihrer Gestaltung*, pp. 81–96.

‡ *Ueber eine neue Art der Riemann'schen Flächen*, pp. 558–566.

there are thus two imaginary tangents corresponding to each point of this region, let us imagine it covered with a double leaf; along the curve the two leaves must, of course, be regarded as joined. Thus we obtain a surface which can be considered as a Riemann surface belonging to the curve, each point of the surface corresponding to a single tangent of the curve. Here, then, we have our anchor-ring. If on such a surface we study integrals, they will be of double periodicity, and the true reason is thus disclosed for the connection of elliptic

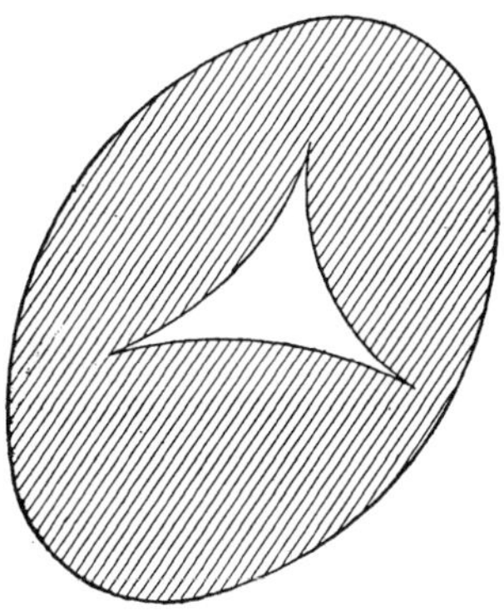

Fig. 5.

integrals with the curves of the third class, and hence, owing to the relation of duality, with the curves of the third order.

To make a further advance, I passed to the general theory of Riemann surfaces. To real curves will of course correspond *symmetrical* Riemann surfaces, *i.e.* surfaces that reproduce themselves by a conformal transformation of the second kind (*i.e.* a transformation that inverts the sense of the angles). Now it is easy to enumerate the different symmetrical types belonging to a given p. The result is that there are altogether $p+1$ "diasymmetric" and $\left[\frac{p+1}{2}\right]$ "orthosymmetric" cases. If we denote as a line of symmetry any line whose points remain unchanged by the conformal transformation, the diasymmetric cases contain respectively p, $p-1, \cdots 2$, 1, 0 lines of symmetry, and the orthosymmetric cases contain $p+1$, $p-1$, $p-3, \cdots$ such lines. A surface is called diasymmetric or orthosymmetric according as it does not or does break up into two parts by cuts carried along all the lines of symmetry. This enumeration, then, will contain a general classification of real curves, as indicated first in my pamphlet on Riemann's theory.* In the summer of 1892 I resumed the theory and developed a large number of propositions concerning the reality of the roots of those equations connected with our curves that can be treated by means of the Abelian integrals. Compare the last volume of the *Math. Annalen*† and my (lithographed) lectures on *Riemann'sche Flächen*, Part II.

In the same manner in which we have to-day considered ordinary algebraic curves and surfaces, it would be interesting to investigate *all* algebraic configurations so as to arrive at a truly geometrical intuition of these objects.

In concluding, I wish to insist in particular on what I regard as the principal characteristic of the geometrical methods that I have discussed to-day: these methods give us an *actual mental image* of the configuration under discussion, and this I consider as most essential in all true geometry. For this reason the so-called synthetic methods, as usually developed, do not appear to me very satisfactory. While giving elaborate constructions for special cases and details they fail entirely to afford a general view of the configurations as a whole.

---

* *Ueber Riemann's Theorie der algebraischen Functionen und ihrer Integrale*, Leipzig, Teubner, 1882. An English translation by Frances Hardcastle (London, Macmillan) has just appeared.

† *Ueber Realitätsverhältnisse bei der einem beliebigen Geschlechte zugehörigen Normalcurve der ϕ*, Vol. 42 (1893), pp. 1–29.

Lecture V.: THEORY OF FUNCTIONS AND GEOMETRY.

(September 1, 1893.)

A geometrical representation of a function of a complex variable $w=f(z)$, where $w=u+iv$ and $z=x+iy$, can be obtained by constructing models of the two surfaces $u=\phi(x, y)$, $v=\psi(x, y)$. This idea is realized in the models constructed by Dyck, which I have shown to you at the Exhibition.

Another well-known method, proposed by Riemann, consists in representing each of the two complex variables in the usual way in a plane. To every point in the z-plane will correspond one or more points in the w-plane; as z moves in its plane, w describes a corresponding curve in the other plane. I may refer to the work of Holzmüller * as a good elementary introduction to this subject, especially on account of the large number of special cases there worked out and illustrated by drawings.

In higher investigations, what is of interest is not so much the corresponding curves as corresponding areas or *regions* of the two planes. According to Riemann's fundamental theorem concerning conformal representation, two simply connected regions can always be made to correspond to each other conformally, so that either is the conformal representation

* *Einführung in die Theorie der isogonalen Verwandtschaften und der conformen Abbildungen, verbunden mit Anwendungen auf mathematische Physik,* Leipzig, Teubner, 1882.

(*Abbildung*) of the other. The three constants at our disposal in this correspondence allow us to select three arbitrary points on the boundary of one region as corresponding to three arbitrary points on the boundary of the other region. Thus Riemann's theory affords a geometrical definition for any function whatever by means of its conformal representation.

This suggests the inquiry as to what conclusions can be drawn from this method concerning the nature of transcendental functions. Next to the elementary transcendental functions the elliptic functions are usually regarded as the most important. There is, however, another class for which at least equal importance must be claimed on account of their numerous applications in astronomy and mathematical physics; these are the *hypergeometric functions*, so called owing to their connection with Gauss's hypergeometric series.

The hypergeometric functions can be defined as the integrals of the following linear differential equation of the second order:

$$\frac{d^2w}{dz^2}+\left[\frac{1-\lambda'-\lambda''}{z-a}(a-b)(a-c)+\frac{1-\mu'-\mu''}{z-b}(b-c)(b-a)\right.$$

$$\left.+\frac{1-\nu'-\nu''}{z-c}(c-a)(c-b)\right]\frac{dw}{dz}+\left[\frac{\lambda'\lambda''(a-b)(a-c)}{z-a}\right.$$

$$\left.+\frac{\mu'\mu''(b-c)(b-a)}{z-b}+\frac{\nu'\nu''(c-a)(c-b)}{z-c}\right]\frac{w}{(z-a)(z-b)(z-c)}=0,$$

where $z=a, b, c$ are the three singular points and λ', λ''; μ', μ''; ν', ν'' are the so-called exponents belonging respectively to a, b, c.

If w_1 be a particular solution, w_2 another, the general solution can be put in the form $\alpha w_1+\beta w_2$, where α, β are arbitrary constants; so that

$$\alpha w_1+\beta w_2 \text{ and } \gamma w_1+\delta w_2$$

represent a pair of general solutions.

If we now introduce the quotient $\frac{w_1}{w_2} = \eta(z)$ as a new variable, its most general value is $\frac{\alpha w_1 + \beta w_2}{\gamma w_1 + \delta w_2} = \frac{\alpha\eta + \beta}{\gamma\eta + \delta}$ and contains therefore three arbitrary constants. Hence η satisfies a differential equation of the third order which is readily found to be

$$\frac{\eta'''}{\eta'} - \frac{3}{2}\left(\frac{\eta''}{\eta'}\right)^2$$

$$= \frac{1}{(z-a)(z-b)(z-c)}\left[\frac{\frac{1-\lambda^2}{2}}{z-a}(a-b)(a-c) + \frac{\frac{1-\mu^2}{2}}{z-b}(b-c)(b-a) + \frac{\frac{1-\nu^2}{2}}{z-c}(c-a)(c-b)\right],$$

in which the left-hand member has the property of not being changed by a linear substitution, and is therefore called a differential invariant. Cayley has named this function the Schwarzian derivative; it has formed the starting-point for Sylvester's investigations on reciprocants. In the right-hand member,

$$\pm\lambda = \lambda' - \lambda'', \ \pm\mu = \mu' - \mu'', \ \pm\nu = \nu' - \nu''.$$

As to the conformal representation (Fig. 6), it can be shown that the upper half of the z-plane, with the points a, b, c on

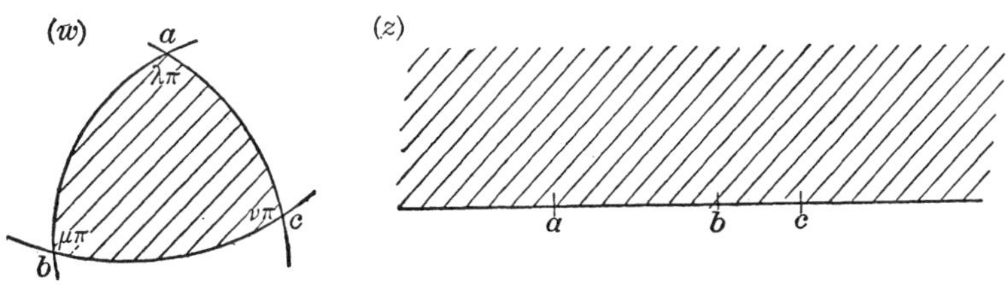

Fig. 6.

the real axis and λ, μ, ν assumed as real, is transformed for each branch of the η-function into a triangular area abc bounded by three circular arcs; let us call such an area a *circular triangle* (*Kreisbogendreieck*). The angles at the vertices of this triangle are $\lambda\pi$, $\mu\pi$, $\nu\pi$.

This, then, is the geometrical representation we have to take as our basis. In order to derive from it conclusions as to the nature of the transcendental functions defined by the differential equation, it will evidently be necessary to inquire what are the forms of such circular triangles in the most general case. For it is to be noticed that there is no restriction laid upon the values of the constants λ, μ, ν, so that the angles of our triangle are not necessarily acute, nor even convex; in other words, in the general case the vertices will be branch-points. The triangle itself is here to be regarded as something like an extensible and flexible membrane spread out between the circles forming the boundary.

I have investigated this question in a paper published in the *Math. Annalen*, Vol. 37.* It will be convenient to project the plane containing the circular triangle stereographically on a sphere. The question then is as to the most general form of spherical triangles, taking this term in a generalized meaning as denoting any triangle on the sphere bounded by the intersections of three planes with the sphere, whether the planes intersect at the centre or not.

This is really a question of elementary geometry; and it is interesting to notice how often in recent times higher research has led back to elementary problems not previously settled.

The result in the present case is that there are two, and only two, species of such generalized triangles. They are obtained from the so-called elementary triangle by two distinct operations: (*a*) *lateral*, (*b*) *polar attachment* of a circle.

---

* *Ueber die Nullstellen der hypergeometrischen Reihe*, pp. 573–590.

Let *abc* (Fig. 7) be the elementary spherical triangle. Then the operation of lateral attachment consists in attaching to the area *abc* the area enclosed by one of the sides, say *bc*, this side being produced so as to form a complete circle. The process can, of course, be repeated any number of times and applied to each side. If one circular area be attached at *bc*, the angles at *b* and *c* are increased each by π; if the whole sphere be attached, by 2π, etc. The vertices in this way become branch-points. A triangle so obtained I call *a triangle of the first species.*

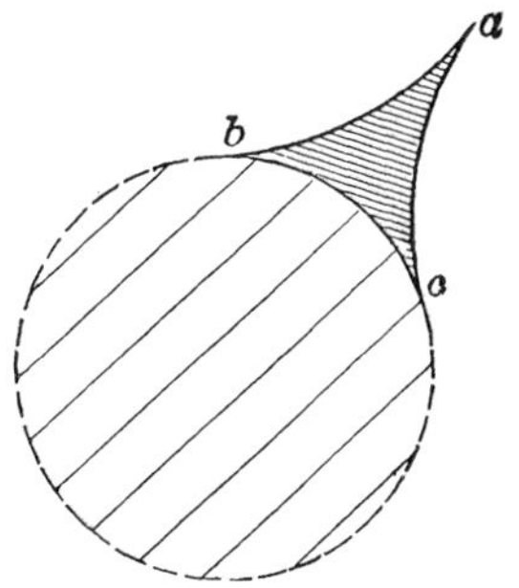

Fig. 7.

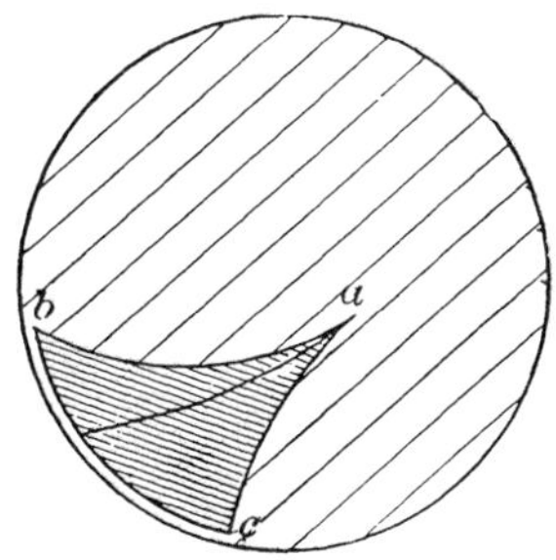

Fig. 8.

A *triangle of the second species* is produced by the process of polar attachment of a circle, say at *bc*; the whole area bounded by the circle *bc* is, in this case, connected with the original triangle along a branch-cut reaching from the vertex *a* to some point on *bc*. The point *a* becomes a branch-point, its angle being increased by 2π. Moreover, lateral attachments can be made at *ab* and *ac*.

The two species of triangles are now characterized as follows: *the first species may have any number of lateral attachments at any or all of the three sides, while the second has a polar attachment to one vertex and the opposite side, and may have lateral attachments to the other two sides.*

Analytically the two species are distinguished by inequalities between the absolute values of the constants λ, μ, ν. For the first species, none of the three constants is greater than the sum of the other two, *i.e.*

$$|\lambda| \leqq |\mu| + |\nu|, \quad |\mu| \leqq |\nu| + |\lambda|, \quad |\nu| \leqq |\lambda| + |\mu|;$$

for the second species,

$$|\lambda| \geqq |\mu| + |\nu|,$$

where λ refers to the pole.

For the application to the theory of functions, it is important to determine, in the case of the second species, the number of times the circle formed by the side opposite the vertex is passed around. I have found this number to be $E\left(\frac{|\lambda| - |\mu| - |\nu| + 1}{2}\right)$, where E denotes the greatest positive integer contained in the argument, and is therefore always zero when this argument happens to be negative or fractional.

Let us now apply these geometrical ideas to the theory of hypergeometric functions. I can here only point out one of the results obtained. Considering only the real values that $\eta = w_1/w_2$ can assume between a and b, the question presents itself as to the shape of the η-curve between these limits. Let us consider for a moment the curves w_1 and w_2. It is well known that, if w_1 oscillates between a and b from one side of the axis to the other, w_2 will also oscillate; their quotient $\eta = w_1/w_2$ is represented by a curve that consists of separate branches extending from $-\infty$ to $+\infty$, somewhat like the curve $y = \tan x$. Now it appears as the result of the investigation that the number of these branches, and therefore the number of the oscillations of w_1 and w_2, is given precisely by the number of circuits of the point c; that is to say, it is $E\left(\frac{|\nu| - |\lambda| - |\mu| + 1}{2}\right)$. This is a result of importance for all applications of hypergeometric functions which was derived only later (by Hurwitz) by means of Sturm's methods.

I wish to call your particular attention not so much to the result itself, however interesting it may be, as to the geometrical method adopted in deriving it. More advanced researches on a similar line of thought are now being carried on at Göttingen by myself and others.

When a differential equation with a larger number of singular points than three is the object of investigation, the triangles must be replaced by quadrangles and other polygons. In my lithographed lectures on *Linear Differential Equations*, delivered in 1890–91, I have thrown out some suggestions regarding the treatment of such cases. The difficulty arising in these generalizations is, strange to say, merely of a geometrical nature, viz. the difficulty of obtaining a general view of the possible forms of the polygons.

Meanwhile, Dr. Schoenflies has published a paper on rectilinear polygons of any number of sides * while Dr. Van Vleck has considered such rectilinear polygons together with the functions they define, the polygons being defined in so general a way as to admit branch-points even in the interior. Dr. Schoenflies has also treated the case of circular quadrangles, the result being somewhat complicated.

In all these investigations the singular points of the z-plane corresponding to the vertices of the polygons are of course assumed to be real, as are also their exponents. There remains the still more general question how to represent by conformal correspondence the functions in the case when some of these elements are complex. In this direction I have to mention the name of Dr. Schilling who has treated the case of the ordinary hypergeometric function on the assumption of complex exponents.

---

* *Ueber Kreisbogenpolygone*, Math. Annalen, Vol. 42, pp. 377–408.

This treatment of the functions defined by linear differential equations of the second order is of course only an example of the general discussion of complex functions by means of geometry. I hope that many more interesting results will be obtained in the future by such geometrical methods.

Lecture VI.: ON THE MATHEMATICAL CHARACTER OF SPACE-INTUITION AND THE RELATION OF PURE MATHEMATICS TO THE APPLIED SCIENCES.

(September 2, 1893.)

In the preceding lectures I have laid so much stress on geometrical methods that the inquiry naturally presents itself as to the real nature and limitations of geometrical intuition.

In my address before the Congress of Mathematics at Chicago I referred to the distinction between what I called the *naïve* and the *refined* intuition. It is the latter that we find in Euclid; he carefully develops his system on the basis of well-formulated axioms, is fully conscious of the necessity of exact proofs, clearly distinguishes between the commensurable and incommensurable, and so forth.

The naïve intuition, on the other hand, was especially active during the period of the genesis of the differential and integral calculus. Thus we see that Newton assumes without hesitation the existence, in every case, of a velocity in a moving point, without troubling himself with the inquiry whether there might not be continuous functions having no derivative.

At the present time we are wont to build up the infinitesimal calculus on a purely analytical basis, and this shows that we are living in a *critical* period similar to that of Euclid. It is my private conviction, although I may perhaps not be able to fully substantiate it with complete proofs, that Euclid's period also must have been preceded by a "naïve" stage of development. Several facts that have become known only quite recently point in this direction. Thus it is now known that the books that have come down to us from the time of Euclid constitute only a very small part of what was then in existence; moreover, much of the teaching was done by oral tradition. Not many of the books had that artistic finish that we admire in Euclid's "Elements"; the majority were in the form of improvised lectures, written out for the use of the students. The investigations of Zeuthen * and Allman † have done much to clear up these historical conditions.

If we now ask how we can account for this distinction between the naïve and refined intuition, I must say that, in my opinion, the root of the matter lies in the fact that *the naïve intuition is not exact, while the refined intuition is not properly intuition at all, but arises through the logical development from axioms considered as perfectly exact.*

To explain the meaning of the first half of this statement it is my opinion that, in our naïve intuition, when thinking of a point we do not picture to our mind an abstract mathematical point, but substitute something concrete for it. In imagining a line, we do not picture to ourselves "length without breadth," but a *strip* of a certain width. Now such a strip has of course *always* a tangent (Fig. 9); *i.e.* we can always imagine a straight strip having a small portion (element) in common with the curved strip; similarly with respect to the osculating circle. The definitions in this case are regarded as holding only approximately, or as far as may be necessary.

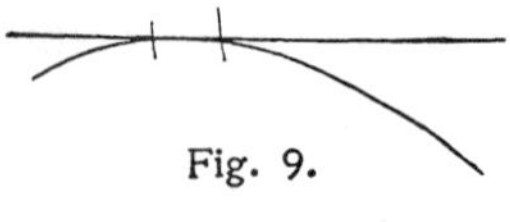

Fig. 9.

* *Die Lehre von den Kegelschnitten im Altertum*, übersetzt von R. v. Fischer-Benzon, Kopenhagen, Höst, 1886.

† *Greek geometry from Thales to Euclid*, Dublin, Hodges, 1889.

The "exact" mathematicians will of course say that such definitions are not definitions at all. But I maintain that in ordinary life we actually operate with such inexact definitions. Thus we speak without hesitancy of the direction and curvature of a river or a road, although the "line" in this case has certainly considerable width.

As regards the second half of my proposition, there actually are many cases where the conclusions derived by purely logical reasoning from exact definitions can no more be verified by intuition. To show this, I select examples from the theory of automorphic functions, because in more common geometrical illustrations our judgment is warped by the familiarity of the ideas.

Let any number of non-intersecting circles 1, 2, 3, 4, ⋯, be given (Fig. 10), and let every circle be reflected (*i.e.* transformed

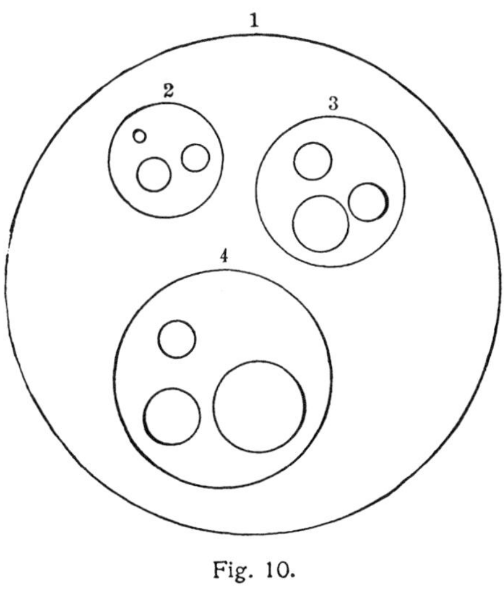

Fig. 10.

by inversion, or reciprocal radii vectores) upon every other circle; then repeat this operation again and again, *ad infinitum.* The question is, what will be the configuration formed by the totality of all the circles, and in particular what will be the position of the limiting points. There is no difficulty in answering these questions by purely logical reasoning; but the imagination seems to fail utterly when we try to form a mental image of the result.

Again, let a series of circles be given, each circle touching the following, while the last touches the first (Fig. 11). Every circle is now reflected upon every other just as in the preceding example, and the process is repeated indefinitely. The special case when the original points of contact happen to lie on a circle

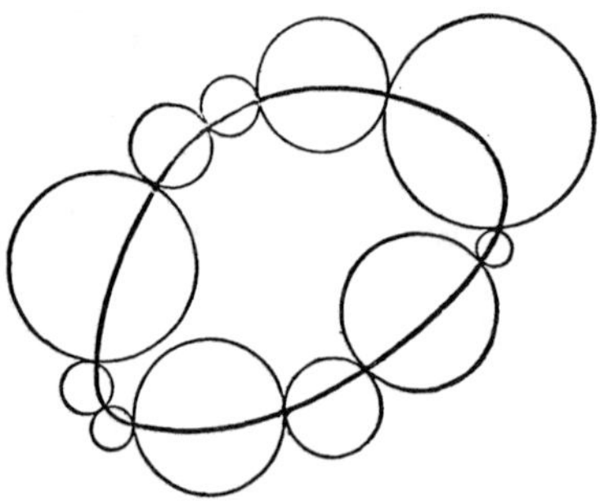

Fig. 11.

being excluded, it can be shown analytically that the continuous curve which is the locus of all the points of contact *is not an analytic curve.* The points of contact form a manifoldness that is everywhere dense on the curve (in the sense of G. Cantor), although there are intermediate points between them. At each of the former points there is a determinate tangent, while there is none at the intermediate points. Second derivatives do not exist at all. It is easy enough to imagine a *strip* covering all these points; but when the width of the strip is reduced beyond a certain limit, we find undulations, and it seems impossible to clearly picture to the mind the final outcome. It is to be noticed that we have here an example of a curve with indeterminate derivatives arising out of purely geometrical considerations, while it might be supposed from the usual treatment of such curves that they can only be defined by artificial analytical series.

Unfortunately, I am not in a position to give a full account of the opinions of philosophers on this subject. As regards the more recent mathematical literature, I have presented my views as developed above in a paper published in 1873, and since reprinted in the *Math. Annalen.** Ideas agreeing in general with mine have been expressed by Pasch, of Giessen, in two works, one on the foundations of geometry,† the other on the principles of the infinitesimal calculus.‡ Another author, Köpcke, of Hamburg, has advanced the idea that our space-intuition is exact as far as it goes, but so limited as to make it impossible for us to picture to ourselves curves without tangents.§

On one point Pasch does not agree with me, and that is as to the exact value of the axioms. He believes — and this is the traditional view — that it is possible finally to discard intuition entirely, basing the whole science on the axioms alone. I am of the opinion that, certainly, for the purposes of research it is always necessary to combine the intuition with the axioms. I do not believe, for instance, that it would have been possible to derive the results discussed in my former lectures, the splendid researches of Lie, the continuity of the shape of algebraic curves and surfaces, or the most general forms of triangles, without the constant use of geometrical intuition.

---

* *Ueber den allgemeinen Functionsbegriff und dessen Darstellung durch eine willkürliche Curve*, Math. Annalen, Vol. 22 (1883), pp. 249–259.

† *Vorlesungen über neuere Geometrie*, Leipzig, Teubner, 1882.

‡ *Einleitung in die Differential- und Integralrechnung*, Leipzig, Teubner, 1882.

§ *Ueber Differentiirbarkeit und Anschaulichkeit der stetigen Functionen*, Math. Annalen, Vol. 29 (1887), pp. 123–140.

Pasch's idea of building up the science purely on the basis of the axioms has since been carried still farther by Peano, in his logical calculus.

Finally, it must be said that the degree of exactness of the intuition of space may be different in different individuals, perhaps even in different races. It would seem as if a strong naïve space-intuition were an attribute pre-eminently of the Teutonic race, while the critical, purely logical sense is more fully developed in the Latin and Hebrew races. A full investigation of this subject, somewhat on the lines suggested by Francis Galton in his researches on heredity, might be interesting.

What has been said above with regard to geometry ranges this science among the applied sciences. A few general remarks on these sciences and their relation to pure mathematics will here not be out of place. From the point of view of pure mathematical science I should lay particular stress on the *heuristic value* of the applied sciences as an aid to discovering new truths in mathematics. Thus I have shown (in my little book on Riemann's theories) that the Abelian integrals can best be understood and illustrated by considering electric currents on closed surfaces. In an analogous way, theorems concerning differential equations can be derived from the consideration of sound-vibrations; and so on.

But just at present I desire to speak of more practical matters, corresponding as it were to what I have said before about the inexactness of geometrical intuition. I believe that the more or less close relation of any applied science to mathematics might be characterized by the degree of exactness attained, or attainable, in its numerical results. Indeed, a rough classification of these sciences could be based simply on the number of significant figures averaged in each. Astronomy (and some branches of physics) would here take the first rank; the number of significant figures attained may here be placed as high as seven, and functions higher than the elementary transcendental functions can be used to advantage. Chemistry would probably be found at the other end of the scale, since in this science rarely more than two or three significant figures can be relied upon. Geometrical drawing, with perhaps 3 to 4 figures, would rank between these extremes; and so we might go on.

The ordinary mathematical treatment of any applied science substitutes exact axioms for the approximate results of experience, and deduces from these axioms the rigid mathematical conclusions. In applying this method it must not be forgotten that mathematical developments transcending the limit of exactness of the science are of no practical value. It follows that a large portion of abstract mathematics remains without finding any practical application, the amount of mathematics that can be usefully employed in any science being in proportion to the degree of accuracy attained in the science. Thus, while the astronomer can put to good use a wide range of mathematical theory, the chemist is only just beginning to apply the first derivative, *i.e.* the rate of change at which certain processes are going on; for second derivatives he does not seem to have found any use as yet.

As examples of extensive mathematical theories that do not exist for applied science, I may mention the distinction between the commensurable and incommensurable, the investigations on the convergency of Fourier's series, the theory of non-analytical functions, etc. It seems to me, therefore, that Kirchhoff makes a mistake when he says in his *Spectral-Analyse* that absorption takes place only when there is *exact* coincidence between the wave-lengths. I side with Stokes, who says that absorption takes place *in the vicinity* of such coincidence. Similarly, when the astronomer says that the periods of two planets must be exactly commensurable to admit the possibility of a collision, this holds only abstractly, for their mathematical centres; and it must be remembered that such things as the period, the mass, etc., of a planet cannot be exactly defined, and are changing all the time. Indeed, we have no way of ascertaining whether two astronomical magnitudes are incommensurable or not; we can only inquire whether their ratio can be expressed approximately by two *small* integers. The statement sometimes made that there exist only analytic functions in nature is in my opinion absurd. All we can say is that we restrict ourselves to analytic, and even only to simple analytic, functions because they afford a sufficient degree of approximation. Indeed, we have the theorem (of Weierstrass) that any continuous function can be approximated to, with any required degree of accuracy, by an analytic function. Thus if $\phi(x)$ be our continuous function, and δ a small quantity representing the given limit of exactness (the width of the strip that we substitute for the curve), it is always possible to determine an *analytic* function $f(x)$ such that

$$\phi(x)=f(x)+\epsilon, \text{ where } |\epsilon|<|\delta|,$$

within the given limits.

All this suggests the question whether it would not be possible to create a, let us say, *abridged* system of mathematics adapted to the needs of the applied sciences, without passing through the whole realm of abstract mathematics. Such a system would have to include, for example, the researches of Gauss on the accuracy of astronomical calculations, or the more recent and highly interesting investigations of Tchebycheff on interpolation. The problem, while perhaps not impossible, seems difficult of solution, mainly on account of the somewhat vague and indefinite character of the questions arising.

I hope that what I have here said concerning the use of mathematics in the applied sciences will not be interpreted as in any way prejudicial to the cultivation of abstract mathematics as a pure science. Apart from the fact that pure mathematics cannot be supplanted by anything else as a means for developing the purely logical powers of the mind, there must be considered here as elsewhere the necessity of the presence of a few individuals in each country developed in a far higher degree than the rest, for the purpose of keeping up and gradually raising the *general* standard. Even a slight raising of the general level can be accomplished only when some few minds have progressed far ahead of the average.

Moreover, the "abridged" system of mathematics referred to above is not yet in existence, and we must for the present deal with the material at hand and try to make the best of it.

Now, just here a practical difficulty presents itself in the teaching of mathematics, let us say of the elements of the differential and integral calculus. The teacher is confronted with the problem of harmonizing two opposite and almost contradictory requirements. On the one hand, he has to consider the limited and as yet undeveloped intellectual grasp of his students and the fact that most of them study mathematics mainly with a view to the practical applications; on the other, his conscientiousness as a teacher and man of science would seem to compel him to detract in nowise from perfect mathematical rigour and therefore to introduce from the beginning all the refinements and niceties of modern abstract mathematics. In recent years the university instruction, at least in Europe, has been tending more and more in the latter direction; and the same tendencies will necessarily manifest themselves in this country in the course of time. The second edition of the *Cours d'analyse* of Camille Jordan may be regarded as an example of this extreme refinement in laying the foundations of the infinitesimal calculus. To place a work of this character in the hands of a beginner must necessarily have the effect that at the beginning a large part of the subject will remain unintelligible, and that, at a later stage, the student will not have gained the power of making use of the principles in the simple cases occurring in the applied sciences.

It is my opinion that in teaching it is not only admissible, but absolutely necessary, to be less abstract at the start, to have constant regard to the applications, and to refer to the refinements only gradually as the student becomes able to understand them. This is, of course, nothing but a universal pedagogical principle to be observed in all mathematical instruction.

Among recent German works I may recommend for the use of beginners, for instance, Kiepert's new and revised edition of Stegemann's text-book;* this work seems to combine simplicity and clearness with sufficient mathematical rigour. On the other hand, it is a matter of course that for more advanced students, especially for professional mathematicians, the study of works like that of Jordan is quite indispensable.

I am led to these remarks by the consciousness of a growing danger in the higher educational system of Germany, — the danger of a separation between abstract mathematical science and its scientific and technical applications. Such separation could only be deplored; for it would necessarily be followed by shallowness on the side of the applied sciences, and by isolation on the part of pure mathematics.

---

* *Grundriss der Differential- und Integral-Rechnung*, 6te Auflage, herausgegeben von Kiepert, Hannover, Helwing, 1892.

LECTURE VII.: THE TRANSCENDENCY OF THE NUMBERS e AND π.

(September 4, 1893.)

LAST Saturday we discussed inexact mathematics; to-day we shall speak of the most exact branch of mathematical science.

It has been shown by G. Cantor that there are two kinds of infinite manifoldnesses: (*a*) *countable* (*abzählbare*) manifoldnesses, whose quantities can be numbered or enumerated so that to each quantity a definite place can be assigned in the system; and (*b*) *non-countable* manifoldnesses, for which this is not possible. To the former group belong not only the rational numbers, but also the so-called *algebraic* numbers, *i.e.* all numbers defined by an algebraic equation,

$$a + a_1x + a_2x^2 + \cdots + a_nx^n = 0$$

with integral coefficients (n being of course a positive integer). As an example of a non-countable manifoldness I may mention the totality of all numbers contained in a *continuum*, such as that formed by the points of the segment of a straight line. Such a continuum contains not only the rational and algebraic numbers, but also the so-called transcendental numbers. The actual existence of transcendental numbers which thus naturally follows from Cantor's theory of manifoldnesses had been proved before, from considerations of a different order, by Liouville. With this, however, is not yet given any means for deciding whether any particular number is transcendental or not. But during the last twenty years it has been established that the two fundamental numbers e and π are really transcendental. It is my object to-day to give you a clear idea of the very simple proof recently given by Hilbert for the transcendency of these two numbers.

The history of this problem is short. Twenty years ago, Hermite * first established the transcendency of e; *i.e.* he showed, by somewhat complicated methods, that the number e cannot be the root of an algebraic equation with integral coefficients. Nine years later, Lindemann,† taking the developments of Hermite as his point of departure, succeeded in proving the transcendency of π. Lindemann's work was verified soon after by Weierstrass.

The proof that π is a transcendental number will forever mark an epoch in mathematical science. It gives the final answer to the problem of squaring the circle and settles this vexed question once for all. This problem requires to derive the number π by a finite number of elementary geometrical processes, *i.e.* with the use of the ruler and compasses alone. As a straight line and a circle, or two circles, have only two intersections, these processes, or any finite combination of them, can be expressed algebraically in a comparatively simple form, so that a solution of the problem of squaring the circle would mean that π can be expressed as the root of an algebraic equation of a comparatively simple kind, viz. one that is solvable by square roots. Lindemann's proof shows that π is not the root of any algebraic equation.

The proof of the transcendency of π will hardly diminish the number of circle-squarers, however; for this class of people has always shown an absolute distrust of mathematicians and a

---

* *Comptes rendus*, Vol. 77 (1873), p. 18, etc.

† *Math. Annalen*, Vol. 20 (1882), p. 213.

contempt for mathematics that cannot be overcome by any amount of demonstration. But Hilbert's simple proof will surely be appreciated by all those who take interest in the establishment of mathematical truths of fundamental importance. This demonstration, which includes the case of the number e as well as that of π, was published quite recently in the *Göttinger Nachrichten.** Immediately after† Hurwitz published a proof for the transcendency of e based on still more elementary principles ; and finally, Gordan ‡ gave a further simplification. All three of these papers will be reprinted in the next *Heft* of the *Math. Annalen.*§ The problem has thus been reduced to such simple terms that the proofs for the transcendency of e and π should henceforth be introduced into university teaching everywhere.

Hilbert's demonstration is based on two propositions. One of these simply asserts the transcendency of e, i.e. *the impossibility of an equation of the form*

$$a + a_1 e + a_2 e^2 + \cdots + a_n e^n = 0, \tag{1}$$

where $a, a_1, a_2, \ldots a_n$ are integral numbers. This is the original proposition of Hermite. To prove the transcendency of π, another proposition (originally due to Lindemann) is required, which asserts *the impossibility of an equation of the form*

$$a + e^{\beta_1} + e^{\beta_2} + \cdots + e^{\beta_n} = 0, \tag{2}$$

where a is an integer, and the exponents are algebraic numbers, viz. the roots of an algebraic equation

$$b\beta^m + b_1\beta^{m-1} + b_2\beta^{m-2} + \cdots + b_m = 0,$$

$b, b_1, b_2, \ldots b_m$ being integers.

---

* 1893, No. 2, p. 113.

† *Ib.*, No. 4.

‡ *Comptes rendus*, 1893, p. 1040.

§ Vol. 43 (1894), pp. 216–224.

It will be noticed that the latter proposition really includes the former as a special case; for it is of course possible that the β's are rational integral numbers, and whenever some of the roots of the equation for β are equal, the corresponding terms in the equation (2) will combine into a single term of the form $a_k e^{\beta_k}$. The former proposition is therefore introduced only for the sake of simplicity.

The central idea of the proof of the impossibility of equation (1) consists in introducing for the quantities $1 : e : e^2 : \ldots : e^n$, in which the equation is homogeneous, proportional quantities

$$I_0 + \epsilon_0 : I_1 + \epsilon_1 : I_2 + \epsilon_2 : \cdots : I_n + \epsilon_n,$$

selected so that each consists of an integer I and a very small fraction ϵ. The equation then assumes the form

$$(aI_0 + a_1 I_1 + \cdots + a_n I_n) + (a\epsilon_0 + a_1\epsilon_1 + \cdots + a_n\epsilon_n) = 0, \tag{3}$$

and it can be shown that the I's and ϵ's can always be so selected as to make the quantity in the first parenthesis, which is of course integral, different from zero, while the quantity in the second parenthesis becomes a proper fraction. Now, as the sum of an integer and a proper fraction cannot be equal to zero, the equation (1) is proved to be impossible.

So much for the general idea of Hilbert's proof. It will be seen that the main difficulty lies in the proper determination of the integers I and the fractions ϵ. For this purpose Hilbert makes use of a definite integral suggested by the investigations of Hermite, viz. the integral

$$J = \int_0^\infty z^\rho [(z-1)\cdots(z-n)]^{\rho+1} e^{-z} dz,$$

where ρ is an integer to be determined afterwards. Multiplying equation (1) term for term by this integral and dividing by $\rho!$, this equation can evidently be put into the form

$$\left(a\frac{\int_0^\infty}{\rho!}+a_1e\frac{\int_1^\infty}{\rho!}+a_2e^2\frac{\int_2^\infty}{\rho!}+\cdots+a_ne^n\frac{\int_n^\infty}{\rho!}\right)$$

$$+\left(a_1e\frac{\int_0^1}{\rho!}+a_2e^2\frac{\int_0^2}{\rho!}+\cdots+a_ne^n\frac{\int_0^n}{\rho!}\right)=0,$$

or designating for shortness the quantities in the two parentheses by P_1 and P_2, respectively,

$$P_1+P_2=0.$$

Now it can be proved that the coefficients of $a, a_1, a_2, \cdots a_n$ in P_1 are all integers, that ρ can be so selected as to make P_1 different from zero, and that at the same time ρ can be taken so large as to make P_2 as small as we please. Thus, equation (1) will be reduced to the impossible form (3).

We proceed to prove these properties of P_1 and P_2. The integral J is readily seen to be an integer divisible by $\rho!$, owing to the well-known relation $\int_0^\infty z^\rho e^{-z}dz=\rho!$. Similarly, by substituting $z=z'+1$, $z=z'+2, \cdots z=z'+n$, it can be shown that $e\int_1^\infty, e^2\int_2^\infty, \dots e^n\int_n^\infty$ are integers divisible by $(\rho+1)!$. It follows that P_1 is an integer, viz.

$$P_1\equiv\pm a(n!)^{\rho+1} \quad [\text{mod.}(\rho+1)].$$

If, therefore, ρ be selected so as to make the right-hand member of this congruence not divisible by $\rho+1$, the whole expression P_1 is different from zero.

As regards the condition that P_2 should be made as small as we please, it can evidently be fulfilled by selecting a sufficiently large value for ρ; this is of course consistent with the condition of making J not divisible by $\rho+1$. For by the theorem of mean values (*Mittelwertsatz*) the integrals can be replaced by powers of constant quantities with ρ in the exponent; and the rate of increase of a power is, for sufficiently large values of ρ, always smaller than that of the factorial which occurs in the denominator.

The proof of the impossibility of equation (2) proceeds on precisely analogous lines. Instead of the integral J we have now to use the integral

$$J' = b^{m(\rho+1)}\int_0^\infty z^\rho[(z-\beta_1)(z-\beta_2)\cdots(z-\beta_m)]^{\rho+1}e^{-z}dz,$$

the β's being the roots of the algebraic equation

$$b\beta^m + b_1\beta^{m-1} + \cdots + b_m = 0.$$

This integral is decomposed as follows:

$$\int_0^\infty = \int_0^\beta + \int_\beta^\infty,$$

where of course the path of integration must be properly determined for complex values of β. For the details I must refer you to Hilbert's paper.

Assuming the impossibility of equation (2), the transcendency of π follows easily from the following considerations, originally given by Lindemann. We notice first, as a consequence of our theorem, that, *with the exception of the point $x=0$, $y=1$, the exponential curve $y=e^x$ has no algebraic point, i.e.* no point both of whose co-ordinates are algebraic numbers. In other words, however densely the plane may be covered with algebraic points, the exponential curve (Fig. 12) manages to pass along the plane without meeting them, the single point (0, 1) excepted. This curious result can be deduced as follows from the impossibility of equation (2). Let y be any algebraic

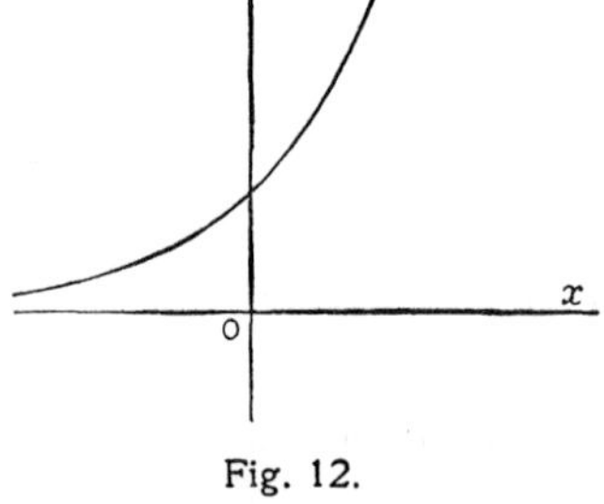

Fig. 12.

quantity, *i.e.* a root of any algebraic equation, and let y_1, y_2, $\cdots$ be the other roots of the same equation; let a similar notation be used for x. Then, if the exponential curve have any algebraic point (x, y), (besides $x=0$, $y=1$), the equation

$$\left.\begin{array}{l}(y-e^{x})(y_1-e^{x})(y_2-e^{x}) \quad \cdots \\ (y-e^{x_1})(y_1-e^{x_1})(y_2-e^{x_1}) \cdots \\ (y-e^{x_2})(y_1-e^{x_2})(y_2-e^{x_2}) \cdots \\ \cdot \quad \cdot \quad \cdot \quad \cdot \quad \cdot \quad \cdot \quad \cdot \quad \cdot \quad \cdot \quad \cdot \quad \cdot \quad \cdot \quad \cdot \quad \cdot \quad \cdot \end{array}\right\} = 0$$

must evidently be fulfilled. But this equation, when multiplied out, has the form of equation (2), which has been shown to be impossible.

As second step we have only to apply the well-known identity

$$1 = e^{i\pi},$$

which is a special case of $y=e^x$. Since in this identity $y=1$ is algebraic, $x=i\pi$ must be transcendental.

LECTURE VIII.: IDEAL NUMBERS.

(September 5, 1893.)

THE theory of numbers is commonly regarded as something exceedingly difficult and abstruse, and as having hardly any connection with the other branches of mathematical science. This view is no doubt due largely to the method of treatment adopted in such works as those of Kummer, Kronecker, Dedekind, and others who have, in the past, most contributed to the advancement of this science. Thus Kummer is reported as having spoken of the theory of numbers as the only *pure* branch of mathematics not yet sullied by contact with the applications.

Recent investigations, however, have made it clear that there exists a very intimate correlation between the theory of numbers and other departments of mathematics, not excluding geometry.

As an example I may mention the theory of the reduction of binary quadratic forms as treated in the *Elliptische Modulfunctionen.* An extension of this method to higher dimensions is possible without serious difficulties. Another example you will remember from the paper by Minkowski, *Ueber Eigenschaften von ganzen Zahlen, die durch räumliche Anschauung erschlossen sind,* which I had the pleasure of presenting to you in abstract at the Congress of Mathematics. Here geometry is used directly for the development of new arithmetical ideas.

To-day I wish to speak on the *composition of binary algebraic forms,* a subject first discussed by Gauss in his *Disquisitiones arithmeticæ** and of Kummer's corresponding theory of *ideal numbers.* Both these subjects have always been considered as very abstruse, although Dirichlet has somewhat simplified the treatment of Gauss. I trust you will find that the geometrical considerations by means of which I shall treat these questions introduce so high a degree of simplicity and clearness that for those not familiar with the older treatment it must be difficult to realize why the subject should ever have been regarded as so very intricate. These considerations were indicated by myself in the *Göttinger Nachrichten* for January, 1893; and at the beginning of the summer semester of the present year I treated them in more extended form in a course of lectures. I have since learned that similar ideas were proposed by Poincaré in 1881; but I have not yet had sufficient leisure to make a comparison of his work with my own.

I write a binary quadratic form as follows:

$$f = ax^2 + bxy + cy^2,$$

i.e. without the factor 2 in the second term; some advantages of this notation were recently pointed out by H. Weber, in the *Göttinger Nachrichten,* 1892–93. The quantities a, b, c, x, y are here of course all assumed to be integers.

It is to be noticed that in the theory of numbers a common factor of the coefficients a, b, c cannot be introduced or omitted arbitrarily, as in projective geometry; in other words, we are concerned with the form, not with an equation. Hence we make the supposition that the coefficients a, b, c have no common factor; a form of this character is called a *primitive form.*

* In the 5th section; see Gauss's *Werke,* Vol. I, p. 239.

As regards the discriminant

$$D = b^2 - 4ac,$$

we shall assume that it has no quadratic divisor (and hence cannot be itself a square), and that it is different from zero. Thus D is either $\equiv 0$ or $\equiv 1 \pmod{4}$. Of the two cases,

$$D < 0 \text{ and } D > 0,$$

which have to be considered separately, I select the former as being more simple. Both cases were treated in my lectures referred to before.

The following elementary geometrical interpretation of the binary quadratic form was given by Gauss, who was much inclined to using geometrical considerations in all branches of mathematics. Construct a parallelogram (Fig. 13) with two

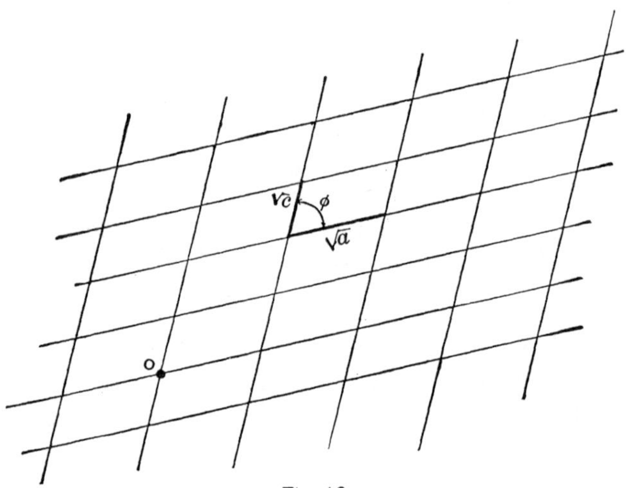

Fig. 13.

adjacent sides equal to $\sqrt{a}$, $\sqrt{c}$, respectively, and the included angle ϕ such that $\cos\phi = \frac{b}{2\sqrt{ac}}$. As $b^2 - 4ac < 0$, a and c have necessarily the same sign; we here assume that a and c are both positive; the case when they are both negative can readily be treated by changing the signs throughout. Next produce the sides of the parallelogram indefinitely, and draw parallels so as to cover the whole plane by a network of equal parallelograms. I shall call this a *line-lattice* (*Parallel-gitter*).

We now select any one of the intersections, or *vertices*, as origin O, and denote every other vertex by the symbol (x, y), x being the number of sides $\sqrt{a}$, y that of sides $\sqrt{c}$, which must be traversed in passing from O to (x, y). Then every value that the form f takes for integral values of x, y evidently represents the square of the distance of the point (x, y) from O. Thus the lattice gives a complete geometrical representation of the binary quadratic form. The discriminant D has also a simple geometrical interpretation, the area of each parallelogram being $=\frac{1}{2}\sqrt{-D}$.

Now, in the theory of numbers, two forms

$$f=ax^2+bxy+cy^2 \quad \text{and} \quad f'=a'x'^2+b'x'y'+c'y'^2$$

are regarded as equivalent if one can be derived from the other by a linear substitution whose determinant is 1, say

$$x'=\alpha x+\beta y, \quad y'=\gamma x+\delta y,$$

where $\alpha\delta-\beta\gamma=1$, α, β, γ, δ being integers. All forms equivalent to a given one are said to compose a *class* of quadratic forms; these forms have all the same discriminant. What corresponds to this equivalence in our geometrical representation will readily appear if we fix our attention on the vertices only (Fig. 14); we then obtain what I propose to call a *point-lattice* (*Punktgitter*). Such a network of points can be connected in various ways by two sets of parallel lines; *i.e.* the point-lattice represents an infinite number of line-lattices. Now it results from an elementary investigation that the point-lattice is the geometrical image of the *class* of binary quadratic forms, the infinite number of line-lattices contained in the point-lattice corresponding exactly to the infinite number of binary forms contained in the class.

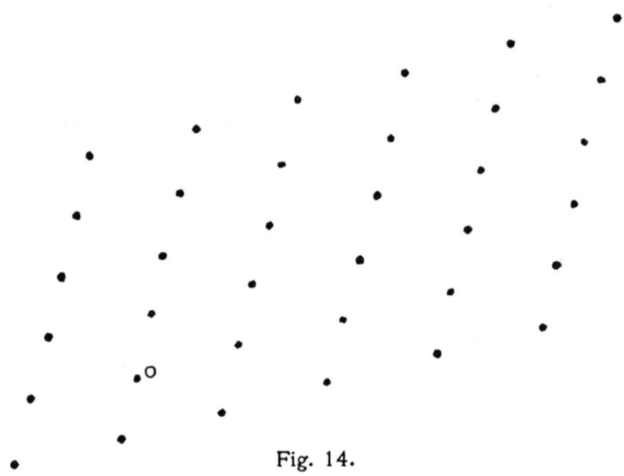

Fig. 14.

It is further known from the theory of numbers that to every value of D belongs only a finite number of classes; hence to every D will correspond a finite number of point-lattices, which we shall afterwards consider together.

Among the different classes belonging to the same value of D, there is one class of particular importance, which I call the *principal class.* It is defined as containing the form

$$x^2 - \tfrac{1}{4}Dy^2$$

when $D \equiv 0 \pmod{4}$, and the form

$$x^2 + xy + \tfrac{1}{4}(1 - D)y^2,$$

when $D \equiv 1 \pmod{4}$. It is easy to see that the corresponding lattices are very simple. When $D \equiv 0 \pmod{4}$, the principal lattice is rectangular, the sides of the elementary parallelogram being 1 and $\sqrt{-\frac{1}{4}D}$. For $D \equiv 1 \pmod{4}$, the parallelogram becomes a rhombus. For the sake of simplicity, I shall here consider only the former case.

Let us now define complex numbers in connection with the principal lattice of the rectangular type (Fig. 15). The point

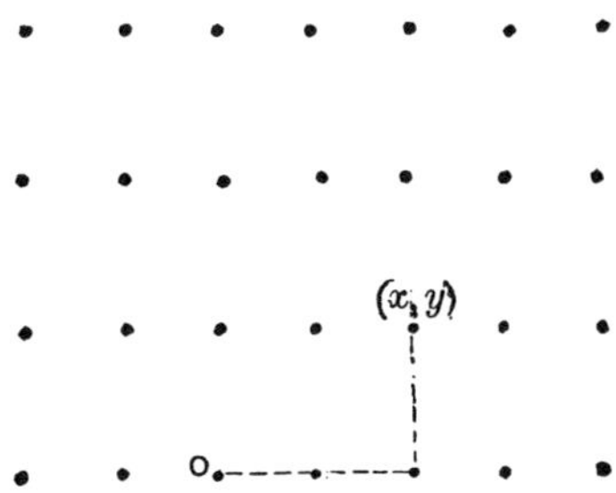

Fig. 15.

(x, y) of the lattice will represent simply the complex number

$$x+\sqrt{-\tfrac{1}{4}D}\cdot y;$$

such numbers we shall call *principal numbers.*

In any system of numbers the laws of multiplication are of prime importance. For our principal numbers it is easy to prove that the product of any two of them always gives a principal number; i.e. *the system of principal numbers is, for multiplication, complete in itself.*

We proceed next to the consideration of lattices of discriminant D that do not belong to the principal class; let us call them *secondary lattices* (*Nebengitter*). Before investigating the laws of multiplication of the corresponding numbers, I must call attention to the fact that there is one feature of arbitrariness in our representation that has not yet been taken into account; this is the *orientation* of the lattice, which may be regarded as given by the angles, ψ and χ, made by the sides

$\sqrt{a}$, $\sqrt{c}$, respectively, with some fixed initial line (Fig. 16). For the angle ϕ of the parallelogram we have evidently $\phi = \chi - \psi$. The point (x, y) of the lattice will thus give the complex number

$$e^{i\psi}\left[\sqrt{a}\cdot x + \frac{-b+\sqrt{D}}{2\sqrt{a}}\cdot y\right] = e^{i\psi}\cdot\sqrt{a}\cdot x + e^{i\chi}\cdot\sqrt{c}\cdot y,$$

which we call a *secondary number*. The definition of a secondary number is therefore indeterminate as long as ψ or χ is not fixed.

Now, by determining ψ properly for every secondary point-lattice, it is always possible to bring about the important result

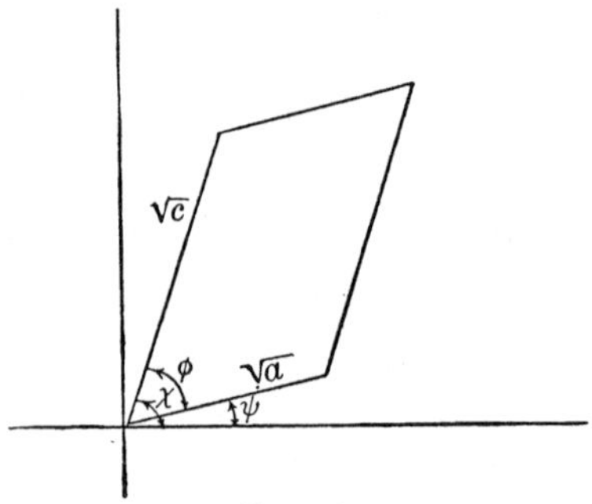

Fig. 16.

that *the product of any two complex numbers of all our lattices taken together will again be a complex number of the system*, so that the totality of these complex numbers forms, likewise, for multiplication, a complete system.

Moreover, the multiplication combines the lattices in a definite way; thus, if any number belonging to the lattice L_1 be multiplied into any number of the lattice L_2, we always obtain a number belonging to a definite lattice L_3.

These properties will be seen to correspond exactly to the characteristic properties of Gauss's *composition of algebraic forms*. For Gauss's law merely asserts that the product of two ordinary numbers that can be represented by two primitive forms f_1, f_2 of discriminant D is always representable by a definite primitive form f_3 of discriminant D. This law is included in the theorem just stated, inasmuch as the values of $\sqrt{f_1}$, $\sqrt{f_2}$, $\sqrt{f_3}$ represent the distances of the points in the lattices from the origin. At the same time we notice that Gauss's law is not exactly equivalent to our theorem, since in the multiplication of our complex numbers, not only the distances are multiplied, but the angles ϕ are added.

It is not impossible that Gauss himself made use of similar considerations in deducing his law, which, taken apart from this geometrical illustration, bears such an abstruse character.

It now remains to explain what relation these investigations have to the ideal numbers of Kummer. This involves the question as to the division of our complex numbers and their resolution into primes.

In the ordinary theory of real numbers, every number can be resolved into primes in only one way. Does this fundamental law hold for our complex numbers? In answering this question we must distinguish between the system formed by the totality of all our complex numbers and the system of principal numbers alone. For the former system the answer is: yes, every complex number can be decomposed into complex primes in only one way. We shall not stop to consider the proof which is directly contained in the ordinary theory of binary quadratic forms. But if we proceed to the consideration of the system of principal numbers alone, the matter is different. There are cases when a principal number can be decomposed in more than one way into prime factors, *i.e.* principal numbers not decomposable into principal factors. Thus it may happen that we have $m_1 m_2 = n_1 n_2$; m_1, m_2, n_1, n_2 being principal primes. The reason is, that these principal numbers are no longer primes if we adjoin the secondary numbers, but are decomposable as follows:

$$m_1 = \alpha \cdot \beta, \quad m_2 = \gamma \cdot \delta,$$
$$n_1 = \alpha \cdot \gamma, \quad n_2 = \beta \cdot \delta,$$

$\alpha, \beta, \gamma, \delta$ being primes in the enlarged system. *In investigating the laws of division it is therefore not convenient to consider the principal system by itself; it is best to introduce the secondary systems.* Kummer, in studying these questions, had originally at his disposal only the principal system; and noticing the imperfection of the resulting laws of division, he introduced by definition his *ideal* numbers so as to re-establish the ordinary laws of division. These ideal numbers of Kummer are thus seen to be nothing but abstract representatives of our secondary numbers. The whole difficulty encountered by every one when first attacking the study of Kummer's ideal numbers is therefore merely a result of his mode of presentation. By introducing from the beginning the secondary numbers by the side of the principal numbers, no difficulty arises at all.

It is true that we have here spoken only of complex numbers containing square roots, while the researches of Kummer himself and of his followers, Kronecker and Dedekind, embrace all possible algebraic numbers. But our methods are of universal application; it is only necessary to construct lattices in spaces of higher dimensions. It would carry us too far to enter into details.

Lecture IX.: THE SOLUTION OF HIGHER ALGEBRAIC EQUATIONS.

(September 6, 1893.)

Formerly the "solution of an algebraic equation" used to mean its solution by radicals. All equations whose solutions cannot be expressed by radicals were classed simply as *insoluble*, although it is well known that the Galois groups belonging to such equations may be very different in character. Even at the present time such ideas are still sometimes found prevailing; and yet, ever since the year 1858, a very different point of view should have been adopted. This is the year in which Hermite and Kronecker, together with Brioschi, found the solution of the equation of the fifth degree, at least in its fundamental ideas.

This solution of the quintic equation is often referred to as a "solution by elliptic functions"; but this expression is not accurate, at least not as a counterpart to the "solution by radicals." Indeed, the elliptic functions enter into the solution of the equation of the fifth degree, as logarithms might be said to enter into the solution of an equation by radicals, because the radicals can be computed by means of logarithms. *The solution of an equation will*, in the present lecture, *be regarded as consisting in its reduction to certain algebraic normal equations.* That the irrationalities involved in the latter can, in the case of the quintic equation, be computed by means of tables of elliptic functions (provided that the proper tables of the corresponding class of elliptic functions were available) is an additional point interesting enough in itself, but not to be considered by us to-day.

I have simplified the solution of the quintic, and think that I have reduced it to the simplest form, by introducing the *icosahedron equation* as the proper normal equation.* In other words, the icosahedron equation determines the typical irrationality to which the solution of the equation of the fifth degree can be reduced. This method is capable of being so generalized as to embrace a whole theory of the solution of higher algebraic equations; and to this I wish to devote the present lecture.

It may be well to state that I speak here of equations with coefficients that are not fixed numerically; the equations are considered from the point of view of the theory of functions, the coefficients corresponding to the independent variables.

In saying that an equation is solvable by radicals we mean that it is reducible by algebraic processes to so-called pure equations,

$$\eta^n = z,$$

where z is a known quantity; then only the new question arises, how $\eta = \sqrt[n]{z}$ can be computed. Let us compare from this point of view the icosahedron equation with the pure equation.

The icosahedron equation is the following equation of the 60th degree:

$$\frac{H^3(\eta)}{1728 f^5(\eta)} = z,$$

where H is a numerical expression of the 20th, f one of the 12th degree, while z is a known quantity. For the actual

* See my work *Vorlesungen über das Ikosaeder und die Auflösung der Gleichungen vom fünften Grade*, Leipzig, Teubner, 1884.

forms of H and f as well as other details I refer you to the *Vorlesungen über das Ikosaeder;* I wish here only to point out the characteristic properties of this equation.

(1) Let η be any one of the roots; then the 60 roots can all be expressed as linear functions of η, with known coefficients, such as for instance,

$$\eta,\ \frac{1}{\eta},\quad \epsilon\eta,\quad \frac{(\epsilon-\epsilon^4)\eta-(\epsilon^2-\epsilon^3)}{(\epsilon^2-\epsilon^3)\eta+(\epsilon-\epsilon^4)},\ \text{etc.},$$

where $\epsilon=e^{\frac{2i\pi}{5}}$. These 60 quantities, then, form a group of 60 linear substitutions.

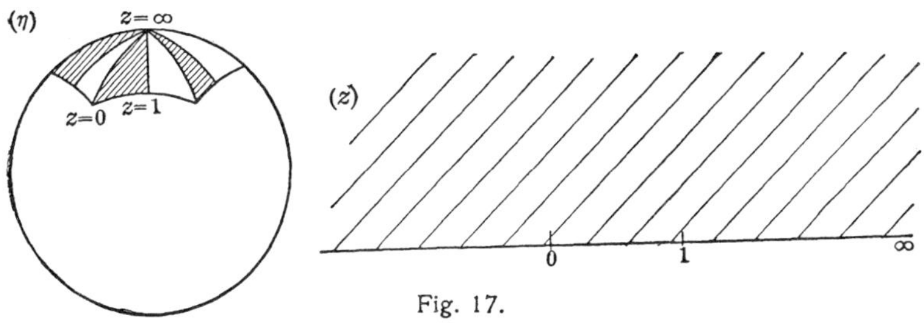

Fig. 17.

(2) Let us next illustrate geometrically the dependence of η on z by establishing the conformal representation of the z-plane on the η-plane, or rather (by stereographic projection) on a sphere (Fig. 17). The triangles corresponding to the upper (shaded) half of the z-plane are the alternate (shaded) triangles on the sphere determined by inscribing a regular icosahedron and dividing each of the 20 triangles so obtained into six equal and symmetrical triangles by drawing the altitudes (Fig. 18). This conformal representation on the sphere assigns to every root a definite region, and is therefore equivalent to a

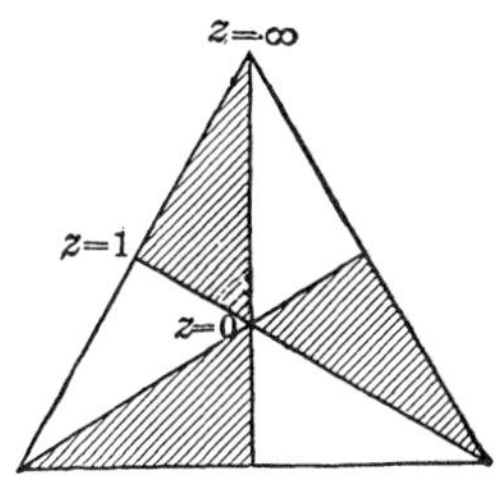

Fig. 18.

perfect separation of the 60 roots. On the other hand, it corresponds in its regular shape to the 60 linear substitutions indicated above.

(3) If, by putting $\eta = y_1/y_2$, we make the 60 expressions of the roots homogeneous, the different values of the quantities y will all be of the form

$$\alpha y_1 + \beta y_2, \ \gamma y_1 + \delta y_2,$$

and therefore satisfy a linear differential equation of the second order

$$y'' + py' + q = 0,$$

p and q being definite rational functions of z. It is, of course, always possible to express every root of an equation by means of a power series. In our case we reduce the calculation of η to that of y_1 and y_2, and try to find series for these quantities. Since these series must satisfy our differential equation of the second order, the law of the series is comparatively simple, any term being expressible by means of the two preceding terms.

(4) Finally, as mentioned before, the calculation of the roots may be abbreviated by the use of elliptic functions, provided tables of such elliptic functions be computed beforehand.

Let us now see what corresponds to each of these four points in the case of the *pure* equation $\eta^n = z$. The results are well known:

(1) All the n roots can be expressed as linear functions of any one of them, η:

$$\eta, \ \epsilon\eta, \ \epsilon^2\eta, \ \cdots \ \epsilon^{n-1}\eta,$$

ϵ being a primitive nth root of unity.

(2) The conformal representation (Fig. 19) gives the division of the sphere into $2n$ equal lunes whose great circles all pass through the same two points.

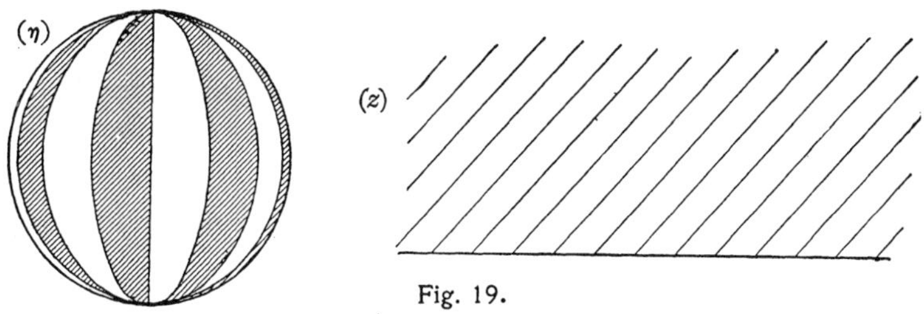

Fig. 19.

(3) There is a differential equation of the first order in η, viz.,

$$nz \cdot \eta' - \eta = 0,$$

from which simple series can be derived for the purposes of actual calculation of the roots.

(4) If these series should be inconvenient, logarithms can be used for computation.

The analogy, you will perceive, is complete. The principal difference between the two cases lies in the fact that, for the pure equation, the linear substitutions involve but one quantity, while for the quintic equation we have a group of *binary* linear substitutions. The same distinction finds expression in the differential equations, the one for the pure equation being of the first order, while that for the quintic is of the second order.

Some remarks may be added concerning the reduction of the general equation of the fifth degree,

$$f_5(x) = 0,$$

to the icosahedron equation. This reduction is possible because the Galois group of our quintic equation (the square root of the discriminant having been adjoined) is isomorphic with the group of the 60 linear substitutions of the icosahedron equation. This possibility of the reduction does not, of course, imply an answer to the question, what operations are needed to effect the reduction. The second part of my *Vorlesungen über das Ikosaeder* is devoted to the latter question. It is found that the reduction cannot be performed rationally, but requires the introduction of a square root. The irrationality thus introduced is, however, an irrationality of a particular kind (a so-called *accessory* irrationality); for it must be such as not to reduce the Galois group of the equation.

I proceed now to consider the general problem of an analogous treatment of higher equations as first given by me in the *Math. Annalen*, Vol. 15 (1879).* I must remark, first of all, that for an accurate exposition it would be necessary to distinguish throughout between the homogeneous and projective formulations (in the latter case, only the ratios of the homogeneous variables are considered). Here it may be allowed to disregard this distinction.

Let us consider the very general problem: *a finite group of homogeneous linear substitutions of* n *variables being given, to calculate the values of the* n *variables from the invariants of the group.*

This problem evidently contains the problem of solving an algebraic equation of any Galois group. For in this case all rational functions of the roots are known that remain unchanged by certain *permutations* of the roots, and permutation is, of course, a simple case of *homogeneous linear transformation.*

Now I propose a general formulation for the treatment of these different problems as follows: *among the problems having isomorphic groups we consider as the simplest the one that has the*

---

* *Ueber die Auflösung gewisser Gleichungen vom siebenten und achten Grade*, pp. 251–282.

least number of variables, and call this the normal problem. This problem must be considered as solvable by series of some kind. The question is to reduce the other isomorphic problems to the normal problem.

This formulation, then, contains what I propose as a general solution of algebraic equations, *i.e.* a reduction of the equations to the isomorphic problem with a minimum number of variables.

The reduction of the equation of the fifth degree to the icosahedron problem is evidently contained in this as a special case, the minimum number of variables being two.

In conclusion I add a brief account showing how far the general problem has been treated for equations of higher degrees.

In the first place, I must here refer to the discussion by myself * and Gordan † of those equations of the seventh degree that have a Galois group of 168 substitutions. The minimum number of variables is here equal to three, the ternary group being the same group of 168 linear substitutions that has since been discussed with full details in Vol. I. of the *Elliptische Modulfunctionen*. While I have confined myself to an exposition of the general idea, Gordan has actually performed the reduction of the equation of the seventh degree to the ternary problem. This is no doubt a splendid piece of work; it is only to be deplored that Gordan here, as elsewhere, has disdained to give his leading ideas apart from the complicated array of formulæ.

Next, I must mention a paper published in Vol. 28 (1887) of the *Math. Annalen*, ‡ where I have shown that for the *general*

* Math. Annalen, Vol. 15 (1879), pp. 251–282.

† *Ueber Gleichungen siebenten Grades mit einer Gruppe von* 168 *Substitutionen*, Math. Annalen, Vol. 20 (1882), pp. 515–530, and Vol. 25 (1885), pp. 459–521.

‡ *Zur Theorie der allgemeinen Gleichungen sechsten und siebenten Grades*, pp. 499–532.

equations of the sixth and seventh degrees the minimum number of the normal problem is four, and how the reduction can be effected.

Finally, in a letter addressed to Camille Jordan* I pointed out the possibility of reducing the equation of the 27th degree, which occurs in the theory of cubic surfaces, to a normal problem containing likewise four variables. This reduction has ultimately been performed in a very simple way by Burkhardt† while all quaternary groups here mentioned have been considered more closely by Maschke.‡

This is the whole account of what has been accomplished; but it is clear that further progress can be made on the same lines without serious difficulty.

A first problem I wish to propose is as follows. In recent years many groups of permutations of 6, 7, 8, 9, ... letters have been made known. The problem would be to determine in each case the minimum number of variables with which isomorphic groups of linear substitutions can be formed.

Secondly, I want to call your particular attention to the case of the general equation of the eighth degree. I have not been able in this case to find a material simplification, so that it would seem as if the equation of the eighth degree were its own normal problem. It would no doubt be interesting to obtain certainty on this point.

---

* Journal de mathématiques, année 1888, p. 169.

† *Untersuchungen aus dem Gebiete der hyperelliptischen Modulfunctionen. Dritter Theil*, Math. Annalen, Vol. 41 (1893), pp. 313–343.

‡ *Ueber die quaternäre, endliche, lineare Substitutionsgruppe der Borchardt'schen Moduln*, Math. Annalen, Vol. 30 (1887), pp. 496–515; *Aufstellung des vollen Formensystems einer quaternären Gruppe von* 51840 *linearen Substitutionen*, ib., Vol. 33 (1889), pp. 317–344; *Ueber eine merkwürdige Configuration gerader Linien im Raume*, ib., Vol. 36 (1890), pp. 190–215.

LECTURE X.: ON SOME RECENT ADVANCES IN HYPERELLIPTIC AND ABELIAN FUNCTIONS.

(September 7, 1893.)

THE subject of hyperelliptic and Abelian functions is of such vast dimensions that it would be impossible to embrace it in its whole extent in one lecture. I wish to speak only of the mutual correlation that has been established between this subject on the one hand, and the theory of invariants, projective geometry, and the theory of groups, on the other. Thus in particular I must omit all mention of the recent attempts to bring arithmetic to bear on these questions. As regards the theory of invariants and projective geometry, their introduction in this domain must be considered as a realization and farther extension of the programme of Clebsch. But the additional idea of groups was necessary for achieving this extension. What I mean by establishing a mutual correlation between these various branches will be best understood if I explain it on the more familiar example of the *elliptic functions*.

To begin with the older method, we have the fundamental elliptic functions in the Jacobian form

$$\sin \operatorname{am}\left(v, \frac{K'}{K}\right), \quad \cos \operatorname{am}\left(v, \frac{K'}{K}\right), \quad \Delta \operatorname{am}\left(v, \frac{K'}{K}\right),$$

as depending on two arguments. These are treated in many works, sometimes more from the geometrical point of view of Riemann, sometimes more from the analytical standpoint of

Weierstrass. I may here mention the first edition of the work of Briot and Bouquet, and of German works those by Königsberger and by Thomae.

The impulse for a new treatment is due to Weierstrass. He introduced, as is well known, three homogeneous arguments, u, ω_1, ω_2, instead of the two Jacobian arguments. This was a necessary preliminary to establishing the connection with the theory of linear substitutions. Let us consider the discontinuous ternary group of linear substitutions,

$$\begin{aligned} u' &= u + m_1\omega_1 + m_2\omega_2, \\ \omega_1' &= \qquad \alpha\omega_1 + \beta\omega_2, \\ \omega_2' &= \qquad \gamma\omega_1 + \delta\omega_2, \end{aligned}$$

where α, β, γ, δ are integers whose determinant $\alpha\delta - \beta\gamma = 1$, while m_1, m_2 are any integers whatever. The fundamental functions of Weierstrass's theory,

$$p(u, \omega_1, \omega_2), \quad p'(u, \omega_1, \omega_2), \quad g_2(\omega_1, \omega_2), \quad g_3(\omega_1, \omega_2),$$

are nothing but the complete system of invariants of that group. It appears, moreover, that g_2, g_3 are also the ordinary (Cayleyan) invariants of the binary biquadratic form $f_4(x_1, x_2)$, on which depends the integral of the first kind

$$\int \frac{x_1 dx_2 - x_2 dx_1}{\sqrt{f_4(x_1, x_2)}}.$$

This significant feature that the transcendental invariants turn out to be at the same time invariants of the algebraic irrationality corresponding to the transcendental theory will hold in all higher cases.

As a next step in the theory of elliptic functions we have to mention the introduction by Clebsch of the systematic consideration of algebraic curves of deficiency 1. He considered in particular the plane curve of the third order (C_3) and the first species of quartic curves (C_4^1) in space, and showed how convenient it is for the derivation of numerous geometrical propositions to regard the elliptic integrals as taken along these curves. The theory of elliptic functions is thus broadened by bringing to bear upon it the ideas of modern projective geometry.

By combining and generalizing these considerations, I was led to the formulation of a very general programme which may be stated as follows (see *Vorlesungen über die Theorie der elliptischen Modulfunctionen*, Vol. II.).

Beginning with the discontinuous group mentioned before

$$\begin{aligned} u' &= u + m_1\omega_1 + m_2\omega_2, \\ \omega_1' &= \alpha\omega_1 + \beta\omega_2, \\ \omega_2' &= \gamma\omega_1 + \delta\omega_2, \end{aligned}$$

our first task is to construct all its sub-groups. Among these the simplest and most useful are those that I have called *congruence sub-groups;* they are obtained by putting

$$\left.\begin{array}{ll} m_1 \equiv 0, & m_2 \equiv 0, \\ \alpha \equiv 1, & \beta \equiv 0, \\ \gamma \equiv 0, & \delta \equiv 1, \end{array}\right\} (\text{mod. } n).$$

The second problem is to construct the invariants of all these groups and the relations between them. Leaving out of consideration all sub-groups except these congruence sub-groups, we have still attained a very considerable enlargement of the theory of elliptic functions. According to the value assigned to the number n, I distinguish different *stages* (*Stufen*) of the problem. It will be noticed that Weierstrass's theory corresponds to the first stage ($n = 1$), while Jacobi's answers, generally speaking, to the second ($n = 2$); the higher stages have not been considered before in a systematic way.

Thirdly, for the purpose of geometrical illustration, I apply Clebsch's idea of the algebraic curve. I begin by introducing the ordinary square root of the binary form which requires the axis of x to be covered twice; *i.e.* we have to use a C_2 in an S_1. I next proceed to the general cubic curve of the plane (C_3 in an S_2), to the quartic curve in space of three dimensions (C_4 in an S_3), and generally to the elliptic curve C_{n+1} in an S_n. These are what I call the normal elliptic curves; they serve best to illustrate any algebraic relations between elliptic functions.

I may notice, by the way, that the treatment here proposed is strictly followed in the *Elliptische Modulfunctionen*, except that there the quantity u is of course assumed to be zero, since this is precisely what characterizes the modular functions. I hope some time to be able to treat the whole theory of elliptic functions (*i.e.* with u different from zero) according to this programme.

The successful extension of this programme to the theory of hyperelliptic and Abelian functions is the best proof of its being a real step in advance. I have therefore devoted my efforts for many years to this extension; and in laying before you an account of what has been accomplished in this rather special field, I hope to attract your attention to various lines of research along which new work can be spent to advantage.

As regards the *hyperelliptic functions*, we may premise as a general definition that they are functions of *two* variables u_1, u_2, with *four* periods (while the elliptic functions have *one* variable u, and *two* periods). Without attempting to give an historical account of the development of the theory of hyperelliptic functions, I turn at once to the researches that mark a progress along the lines specified above, beginning with the geometric application of these functions to surfaces in a space of any number of dimensions.

Here we have first the investigation by Rohn of Kummer's surface, the well-known surface of the fourth order, with 16 conical points. I have myself given a report on this work in the *Math. Annalen*, Vol. 27 (1886).* If every mathematician is struck by the beauty and simplicity of the relations developed in the corresponding cases of the elliptic functions (the C_3 in the plane, etc.), the remarkable configurations inscribed and circumscribed to the Kummer surface that have here been developed by Rohn and myself, should not fail to elicit interest.

Further, I have to mention an extensive memoir by Reichardt, published in 1886, in the *Acta Leopoldina*, where the connection between hyperelliptic functions and Kummer's surface is summarized in a convenient and comprehensive form, as an introduction to this branch. The starting-point of the investigation is taken in the theory of line-complexes of the second degree.

Quite recently the French mathematicians have turned their attention to the general question of the representation of surfaces by means of hyperelliptic functions, and a long memoir by Humbert on this subject will be found in the last volume of the *Journal de Mathématiques*.†

I turn now to the abstract theory of hyperelliptic functions. It is well known that Göpel and Rosenhain established that theory in 1847 in a manner closely corresponding to the Jacobian theory of elliptic functions, the integrals

$$u_1 = \int \frac{dx}{\sqrt{f_6(x)}}, \quad u_2 = \int \frac{x\,dx}{\sqrt{f_6(x)}}$$

taking the place of the single elliptic integral u. Here, then, the question arises: what is the relation of the hyperelliptic functions to the invariants of the binary form of the sixth order $f_6(x_1, x_2)$? In the investigation of this question by myself and

* *Ueber Configurationen, welche der Kummer'schen Fläche zugleich eingeschrieben und umgeschrieben sind*, pp. 106–142.

† *Théorie générale des surfaces hyperelliptiques*, année 1893, pp. 29–170.

Burkhardt, published in Vol. 27 (1886)* and Vol. 32 (1888)† of the *Math. Annalen*, we found that the decompositions of the form f_6 into two factors of lower order, $f_6=\phi_1\psi_5=\phi_3\psi_3$, had to be considered. These being, of course, irrational decompositions, the corresponding invariants are irrational; and a study of the theory of such invariants became necessary.

But another new step had to be taken. The hyperelliptic integrals involve the form f_6 under the square root, $\sqrt{f_6(x_1, x_2)}$. The corresponding Riemann surface has, therefore, two leaves connected at six points; and the problem arises of considering binary forms of x_1, x_2 on such a Riemann surface, just as ordinarily functions of x alone are considered thereon. It can be shown that there exists a particular kind of forms called *primeforms*, strictly analogous to the determinant $x_1y_2-x_2y_1$ in the ordinary complex plane. The primeform on the two-leaved Riemann surface, like this determinant in the ordinary theory, has the property of vanishing only when the points (x_1, x_2) and (y_1, y_2) co-incide (on the same leaf). Moreover, the primeform does not become infinite anywhere. The analogy to the determinant $x_1y_2-x_2y_1$ fails only in so far as the primeform is no longer an algebraic but a transcendental form. Still, all algebraic forms on the surface can be decomposed into prime factors. Moreover, these primeforms give the natural means for the construction of the θ-functions. As an intermediate step we have here functions called by me σ-functions in analogy to the σ-functions of Weierstrass's elliptic theory. In the papers referred to (*Math. Annalen*, Vols. 27, 32) all these considerations are, of course, given for the general case of hyperelliptic functions, the irrationality being $\sqrt{f_{2p+2}(x_1, x_2)}$, where f_{2p+2} is a binary form of the order $2p+2$.

---

* *Ueber hyperelliptische Sigmafunctionen*, pp. 431–464.

† pp. 351–380 and 381–442.

Having thus established the connection between the ordinary theory of hyperelliptic functions of $p=2$ and the invariants of the binary sextic, I undertook the systematic development of what I have called, in the case of elliptic functions, the *Stufentheorie*. The lectures I gave on this subject in 1887–88 have been developed very fully by Burkhardt in the *Math. Annalen*, Vol. 35 (1890).*

As regards the first stage, which, owing to the connection with the theory of *rational* invariants and covariants, requires very complicated calculations, the Italian mathematician, Pascal, has made much progress (*Annali di matematica*). In this connection I must refer to the paper by Bolza† in *Math. Annalen*, Vol. 30 (1887), where the question is discussed in how far it is possible to represent the rational invariants of the sextic by means of the zero values of the θ-functions.

For higher stages, in particular stage three, Burkhardt has given very valuable developments in the *Math. Annalen*, Vol. 36 (1890), p. 371; Vol. 38 (1891), p. 161; Vol. 41 (1893), p. 313. He considers, however, only the hyperelliptic modular functions (u_1 and u_2 being assumed to be zero). The final aim, which Burkhardt seems to have attained, although a large amount of numerical calculation remains to be filled in, consists here in establishing the so-called *multiplier-equation* for transformations of the third order. The equation is of the 40th degree; and Burkhardt has given the general law for the formation of the coefficients.

I invite you to compare his treatment with that of Krause in his book *Die Transformation der hyperelliptischen Functionen erster Ordnung*, Leipzig, Teubner, 1886. His investiga-

---

* *Grundzüge einer allgemeinen Systematik der hyperelliptischen Functionen I. Ordnung*, pp. 198–296.

† *Darstellung der rationalen ganzen Invarianten der Binärform sechsten Grades durch die Nullwerthe der zugehörigen θ-Functionen*, pp. 478–495.

tions, based on the general relations between θ-functions, may go farther; but they are carried out from the purely formal point of view, without reference to the theories of invariants, of groups, or other allied topics.

So much as regards hyperelliptic functions. I now proceed to report briefly on the corresponding advances made in the theory of Abelian functions. I give merely a list of papers; they may be classed under three heads:

(1) A *preliminary* question relates to the invariant representation of the integral of the third kind on algebraic curves of higher deficiency. Pick * has considered this problem for plane curves having no singular points. On the other hand, White, in his dissertation,† briefly reported in *Math. Annalen,* Vol. 36 (1890), p. 597, and printed in full in the *Acta Leopoldina,* has treated such curves in space as are the complete intersection of two surfaces and have no singular point. We may here also notice the researches of Pick and Osgood ‡ on the so-called binomial integrals.

(2) An exposition of the general theory of forms on Riemann surfaces of any kind, in particular a definition of the primeform belonging to each surface, was given by myself in Vol. 36 (1890) of the *Math. Annalen.*§ I may add that during the last year this subject was taken up anew and farther developed by Dr. Ritter; see *Göttinger Nachrichten* for 1893, and *Math. Annalen,* Vol. 44. Dr. Ritter considers the algebraic forms as special cases of more general forms, the *multiplicative forms,* and thus takes a real step in advance.

---

* *Zur Theorie der Abel'schen Functionen,* Math. Annalen, Vol. 29 (1887), pp. 259–271.

† *Abel'sche Integrale auf singularitätenfreien, einfach überdeckten, vollständigen Schnittcurven eines beliebig ausgedehnten Raumes,* Halle, 1891, pp. 43–128.

‡ Osgood, *Zur Theorie der zum algebraischen Gebilde* $y^m = R(x)$ *gehörigen Abel'schen Functionen,* Göttingen, 1890, 8vo, 61 pp.

§ *Zur Theorie der Abel'schen Functionen,* pp. 1–83.

(3) Finally, the particular case $p=3$ has been studied on the basis of our programme in various directions. The normal curve for this case is well known to be the plane quartic C_4 whose geometric properties have been investigated by Hesse and others. I found (*Math. Annalen*, Vol. 36) that these geometrical results, though obtained from an entirely different point of view, corresponded exactly to the needs of the Abelian problem, and actually enabled me to define clearly the 64 θ-functions with the aid of the C_4. Here, as elsewhere, there seems to reign a certain pre-established harmony in the development of mathematics, what is required in one line of research being supplied by another line, so that there appears to be a logical necessity in this, independent of our individual disposition.

In this case, also, I have introduced σ-functions in the place of the θ-functions. The coefficients are irrational covariants just as in the case $p=2$. These σ-series have been studied at great length by Pascal in the *Annali di Matematica.* These investigations bear, of course, a close relation to those of Frobenius and Schottky, which only the lack of time prevents me from quoting in detail.

Finally, the recent investigations of an Austrian mathematician, *Wirtinger*, must here be mentioned. First, Wirtinger has established for $p=3$ the analogue to the Kummer surface; this is a manifoldness of three dimensions and the 24th order in an S_7; see *Göttinger Nachrichten* for 1889, and *Wiener Monatshefte*, 1890. Though apparently rather complicated, this manifoldness has some very elegant properties; thus it is transformed into itself by 64 collineations and 64 reciprocations. Next, in Vol. 40 (1892), of the *Math. Annalen*,* Wirtinger has discussed the Abelian functions on the assumption that only

* *Untersuchungen über Abel'sche Functionen vom Geschlechte* 3, pp. 261–312.

rational invariants and covariants of the curve of the fourth order are to be considered; this corresponds to the "first stage" with $p=3$. The investigation is full of new and fruitful ideas.

In concluding, I wish to say that, for the cases $p=2$ and $p=3$, while much still remains to be done, the fundamental difficulties have been overcome. The great problem to be attacked next is that of $p=4$, where the normal curve is of the sixth order in space. It is to be hoped that renewed efforts will result in overcoming all remaining difficulties. Another promising problem presents itself in the field of θ-functions, when the general θ-series are taken as starting-point, and not the algebraic curve. An enormous number of formulæ have there been developed by analysts, and the problem would be to connect these formulæ with clear geometrical conceptions of the various algebraic configurations. I emphasize these special problems because the Abelian functions have always been regarded as one of the most interesting achievements of modern mathematics, so that every advance we make in this theory gives a standard by which we can measure our own efficiency.

LECTURE XI.: THE MOST RECENT RESEARCHES IN NON-EUCLIDEAN GEOMETRY.

(September 8, 1893.)

MY remarks to-day will be confined to the progress of non-Euclidean geometry during the last few years. Before reporting on these latest developments, however, I must briefly summarize what may be regarded as the general state of opinion among mathematicians in this field. There are three points of view from which non-Euclidean geometry has been considered.

(1) First we have the point of view of elementary geometry, of which Lobachevsky and Bolyai themselves are representatives. Both begin with simple geometrical constructions, proceeding just like Euclid, except that they substitute another axiom for the axiom of parallels. Thus they build up a system of non-Euclidean geometry in which the length of the line is infinite, and the "measure of curvature" (to anticipate a term not used by them) is negative. It is, of course, possible by a similar process to obtain the geometry with a positive measure of curvature, first suggested by Riemann; it is only necessary to formulate the axioms so as to make the length of a line finite, whereby the existence of parallels is made impossible.

(2) From the point of view of projective geometry, we begin by establishing the system of projective geometry in the sense of von Staudt, introducing projective co-ordinates, so that straight lines and planes are given by *linear* equations. Cay-

ley's theory of projective measurement leads then directly to the three possible cases of non-Euclidean geometry: hyperbolic, parabolic, and elliptic, according as the measure of curvature k is <0, $=0$, or >0. It is here, of course, essential to adopt the system of von Staudt and not that of Steiner, since the latter defines the anharmonic ratio by means of distances of points, and not by pure projective constructions.

(3) Finally, we have the point of view of Riemann and Helmholtz. Riemann starts with the idea of the element of distance ds, which he assumes to be expressible in the form

$$ds = \sqrt{\Sigma a_{ik} dx_i dx_k}.$$

Helmholtz, in trying to find a reason for this assumption, considers the motions of a rigid body in space, and derives from these the necessity of giving to ds the form indicated. On the other hand, Riemann introduces the fundamental notion of the *measure of curvature of space.*

The idea of a measure of curvature for the case of two variables, *i.e.* for a surface in a three-dimensional space, is due to Gauss, who showed that this is an intrinsic characteristic of the surface quite independent of the higher space in which the surface happens to be situated. This point has given rise to a misunderstanding on the part of many non-Euclidean writers. When Riemann attributes to his space of three dimensions a measure of curvature k, he only wants to say that there exists an invariant of the "form" $\Sigma a_{ik} dx_i dx_k$; he does not mean to imply that the three-dimensional space necessarily exists as a curved space in a space of four dimensions. Similarly, the illustration of a space of constant positive measure of curvature by the familiar example of the sphere is somewhat misleading. Owing to the fact that on the sphere the geodesic lines (great circles) issuing from any point all meet again in another definite point, antipodal, so to speak, to the original point, the existence of such an antipodal point has sometimes been regarded as a necessary consequence of the assumption of a constant positive curvature. The projective theory of non-Euclidean space shows immediately that the existence of an antipodal point, though compatible with the nature of an elliptic space, is not necessary, but that two geodesic lines in such a space may intersect in one point if at all.*

I call attention to these details in order to show that there is some advantage in adopting the second of the three points of view characterized above, although the third is at least equally important. Indeed, our ideas of space come to us through the senses of vision and motion, the "optical properties" of space forming one source, while the "mechanical properties" form another; the former corresponds in a general way to the projective properties, the latter to those discussed by Helmholtz.

As mentioned before, from the point of view of projective geometry, von Staudt's system should be adopted as the basis. It might be argued that von Staudt practically assumes the axiom of parallels (in postulating a one-to-one correspondence between a pencil of lines and a row of points). But I have shown in the *Math. Annalen*† how this apparent difficulty can be overcome by restricting all constructions of von Staudt to a limited portion of space.

I now proceed to give an account of the most recent researches in non-Euclidean geometry made by Lie and myself. Lie published a brief paper on the subject in the *Berichte* of the Saxon Academy (1886), and a more extensive exposition of his views in the same *Berichte* for 1890 and 1891. These

* This theory has also been developed by Newcomb, in the *Journal für reine und angewandte Mathematik*, Vol. 83 (1877), pp. 293–299.

† *Ueber die sogenannte Nicht-Euklidische Geometrie*, Math. Annalen, Vol. 6 (1873), pp. 112–145.

papers contain an application of Lie's theory of continuous groups to the problem formulated by Helmholtz. I have the more pleasure in placing before you the results of Lie's investigations as they are not taken into due account in my paper on the foundations of projective geometry in Vol. 37 of the *Math. Annalen* (1890),* nor in my (lithographed) lectures on non-Euclidean geometry delivered at Göttingen in 1889–90; the last two papers of Lie appeared too late to be considered, while the first had somehow escaped my memory.

I must begin by stating the problem of Helmholtz in modern terminology. The motions of three-dimensional space are ∞^6, and form a group, say G_6. This group is known to have an invariant for any two points p, p', viz. the distance $\Omega(p, p')$ of these points. But the form of this invariant (and generally the form of the group) in terms of the co-ordinates $x_1, x_2, x_3, y_1, y_2, y_3$ of the points is not known *a priori*. The question arises whether the group of motions is fully characterized by these two properties so that none but the Euclidean and the two non-Euclidean systems of geometry are possible.

For illustration Helmholtz made use of the analogous case in two dimensions. Here we have a group of ∞^3 motions; the distance is again an invariant; and yet it is possible to construct a group not belonging to any one of our three systems, as follows.

Let z be a complex variable; the substitution characterizing the group of Euclidean geometry can be written in the well-known form

$$z' = e^{i\phi}z + m + in = (\cos\phi + i\sin\phi)z + m + in.$$

Now modifying this expression by introducing a complex number in the exponent,

$$z' = e^{(a+i)\phi}z + m + in = e^{a\phi}(\cos\phi + i\sin\phi)z + m + in,$$

* *Zur Nicht-Euklidischen Geometrie*, pp. 544–572.

we obtain a group of transformations by which a point (in the simple case $m=0$, $n=0$) would not move about the origin in a circle, but in a logarithmic spiral; and yet this is a group G_3 with three variable parameters m, n, ϕ, having an invariant for every two points, just like the original group. Helmholtz concludes, therefore, that a new condition, that of *monodromy*, must be added to determine our group completely.

I now proceed to the work of Lie. First as to the results: Lie has confirmed those of Helmholtz with the single exception that in space of three dimensions the axiom of monodromy is not needed, but that the groups to be considered are fully determined by the other axioms. As regards the proofs, however, Lie has shown that the considerations of Helmholtz must be supplemented. The matter is this. In keeping one point of space fixed, our G_6 will be reduced to a G_3. Now Helmholtz inquires how the differentials of the lines issuing from the fixed point are transformed by this G_3. For this purpose he writes down the formulæ

$$dx_1' = a_{11}dx_1 + a_{12}dx_2 + a_{13}dx_3,$$
$$dx_2' = a_{21}dx_1 + a_{22}dx_2 + a_{23}dx_3,$$
$$dx_3' = a_{31}dx_1 + a_{32}dx_2 + a_{33}dx_3,$$

and considers the coefficients a_{11}, a_{12}, $\cdots$ a_{33} as depending on three variable parameters. But Lie remarks that this is not sufficiently general. The linear equations given above represent only the first terms of power series, and the possibility must be considered that the three parameters of the group may not all be involved in the linear terms. In order to treat all possible cases, the general developments of Lie's theory of groups must be applied, and this is just what Lie does.

Let me now say a few words on my own recent researches in non-Euclidean geometry which will be found in a paper published in the *Math. Annalen*, Vol. 37 (1890), p. 544. Their result is that our ideas as to non-Euclidean space are still very incomplete. Indeed, all the researches of Riemann, Helmholtz, Lie; consider only a portion of space surrounding the origin; they establish the existence of analytic laws in the vicinity of that point. Now this space can of course be continued, and the question is to see what kind of connection of space may result from this continuation. It is found that there are different possibilities, each of the three geometries giving rise to a series of subdivisions.

To understand better what is meant by these varieties of connection, let us compare the geometry on a sphere with that in the sheaf of lines formed by the diameters of the sphere. Considering each diameter as an infinite line or ray passing through the centre (not a half-ray issuing from the centre), to each line of the sheaf there will correspond two points on the sphere, viz. the two points of intersection of the line with the sphere. We have, therefore, a one-to-two correspondence between the lines of the sheaf and the points of the sphere. Let us now take a small area on the sphere; it is clear that the distance of two points contained in this area is equal to the angle of the corresponding lines of the sheaf. Thus the geometry of points on the sphere and the geometry of lines in the sheaf are identical as far as small regions are concerned, both corresponding to the assumption of a constant positive measure of curvature. A difference appears, however, as soon as we consider the whole closed sphere on the one hand and the complete sheaf on the other. Let us take, for instance, two geodesic lines of the sphere, *i.e.* two great circles, which evidently intersect in two (diametral) points. The corresponding pencils of the sheaf have only *one* straight line in common.

A second example for this distinction occurs in comparing the geometry of the Euclidean plane with the geometry on a closed cylindrical surface. The latter can be developed in the usual way into a strip of the plane bounded by two parallel lines, as will appear from Fig. 20, the arrows indicating that the opposite points of the edges are coincident on the cylindrical surface. We notice at once the difference: while in the plane all geodesic lines are infinite, on the cylinder there is

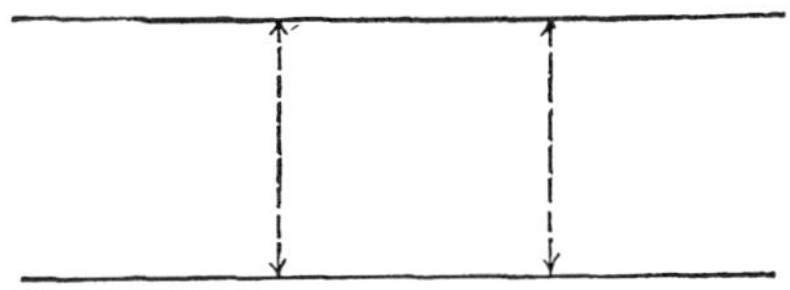

Fig. 20.

one geodesic line that is of finite length, and while in the plane two geodesic lines always intersect in one point, if at all, on the cylinder there may be ∞ points of intersection.

This second example was generalized by Clifford in an address before the Bradford meeting of the British Associa-

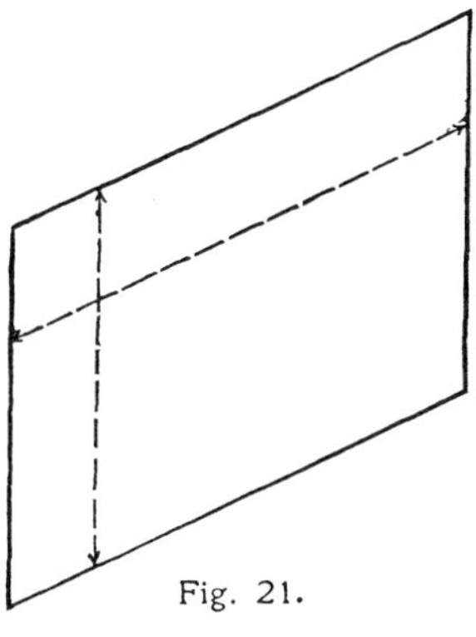

Fig. 21.

tion (1873). In accordance with Clifford's general idea, we may define a closed surface by taking a parallelogram out of an ordinary plane and making the opposite edges correspond point to point as indicated in Fig. 21. It is not to be understood that the opposite edges should be brought to coincidence by bending the parallelogram (which evidently would be impossible without stretching); but only the logical convention is made that the opposite points should be considered as identical. Here, then, we have a closed manifoldness of the connectivity of an anchor-ring, and every one will see the great differences that exist here in comparison with the Euclidean plane in everything concerning the lengths and the intersections of geodesic lines, etc.

It is interesting to consider the G_3 of Euclidean motions on this surface. There is no longer any possibility of moving the surface on itself in ∞^3 ways, the closed surface being considered in its totality. But there is no difficulty in moving any small area over the closed surface in ∞^3 ways.

We have thus found, in addition to the Euclidean plane, two other forms of surfaces: the strip between parallels and Clifford's parallelogram. Similarly we have by the side of ordinary Euclidean space three other types with the Euclidean element of arc; one of these results from considering a parallelepiped.

Here I introduce the axiomatic element. There is no way of proving that the whole of space can be moved in itself in ∞^6 ways; all we know is that small portions of space can be moved in space in ∞^6 ways. Hence there exists the possibility that our actual space, the measure of curvature being taken as zero, may correspond to any one of the four cases.

Carrying out the same considerations for the spaces of constant positive measure of curvature, we are led back to the two cases of elliptic and spherical geometry mentioned before. If, however, the measure of curvature be assumed as a negative constant, we obtain an infinite number of cases, corresponding exactly to the configurations considered by Poincaré and myself in the theory of automorphic functions. This I shall not stop to develop here.

I may add that Killing has verified this whole theory.* It is evident that from this point of view many assertions concerning space made by previous writers are no longer correct (*e.g.* that infinity of space is a consequence of zero curvature), so that we are forced to the opinion that our geometrical demonstrations have no absolute objective truth, but are true only for the present state of our knowledge. These demonstrations are always confined within the range of the space-conceptions that are familiar to us; and we can never tell whether an enlarged conception may not lead to further possibilities that would have to be taken into account. From this point of view we are led in geometry to a certain modesty, such as is always in place in the physical sciences.

---

* *Ueber die Clifford-Klein'schen Raumformen*, Math. Annalen, Vol. 39 (1891), pp. 257–278.

LECTURE XII.: THE STUDY OF MATHEMATICS AT GÖTTINGEN.

(September 9, 1893.)

IN this last lecture I should like to make some general remarks on the way in which the study of mathematics is organized at the university of Göttingen, with particular reference to what may be of interest to American students. At the same time I desire to give you an opportunity to ask any questions that may occur to you as to the broader subject of mathematical study at German universities in general. I shall be glad to answer such inquiries to the extent of my ability.

It is perhaps inexact to speak of an *organization* of the mathematical teaching at Göttingen; you know that *Lern- und Lehr-Freiheit* prevail at a German university, so that the organization I have in mind consists merely in a voluntary agreement among the mathematical professors and instructors. We distinguish at Göttingen between a general and a higher course in mathematics. The general course is intended for that large majority of our students whose intention it is to devote themselves to the teaching of mathematics and physics in the higher schools (*Gymnasien, Realgymnasien, Realschulen*), while the higher course is designed specially for those whose final aim is original investigation.

As regards the former class of students, it is my opinion that in Germany (here in America, I presume, the conditions are very different) the abstractly theoretical instruction given to them has been carried too far. It is no doubt true that what the university should give the student above all other things is the scientific ideal. For this reason even these students should push their mathematical studies far beyond the elementary branches they may have to teach in the future. But the ideal set before them should not be chosen so far distant, and so out of connection with their more immediate wants, as to make it difficult or impossible for them to perceive the bearing that this ideal has on their future work in practical life. In other words, the ideal should be such as to fill the future teacher with enthusiasm for his life-work, not such as to make him look upon this work with contempt as an unworthy drudgery.

For this reason we insist that our students of this class, in addition to their lectures on pure mathematics, should pursue a thorough course in physics, this subject forming an integral part of the curriculum of the higher schools. Astronomy is also recommended as showing an important application of mathematics; and I believe that the technical branches, such as applied mechanics, resistance of materials, etc., would form a valuable aid in showing the practical bearing of mathematical science. Geometrical drawing and descriptive geometry form also a portion of the course. Special exercises in the solution of problems, in lecturing, etc., are arranged in connection with the mathematical lectures, so as to bring the students into personal contact with the instructors.

I wish, however, to speak here more particularly on the higher courses, as these are of more special interest to American students. Here specialization is of course necessary. Each professor and docent delivers certain lectures specially designed for advanced students, in particular for those studying for the doctor's degree. Owing to the wide extent of modern mathematics, it would be out of the question to cover the whole field. These lectures are therefore not regularly repeated every year; they depend largely on the special line of research that happens at the time to engage the attention of the professor. In addition to the lectures we have the higher seminaries, whose principal object is to guide the student in original investigation and give him an opportunity for individual work.

As regards my own higher lectures, I have pursued a certain plan in selecting the subjects for different years, my general aim being *to gain, in the course of time, a complete view of the whole field of modern mathematics, with particular regard to the intuitional or* (in the highest sense of the term) *geometrical standpoint.* This general tendency you will, I trust, also find expressed in this colloquium, in which I have tried to present, within certain limits, a general programme of my individual work. To carry out this plan in Göttingen, and to bring it to the notice of my students, I have, for many years, adopted the method of having my higher lectures carefully written out, and, in recent years, of having them lithographed, so as to make them more readily accessible. These former lectures are at the disposal of my hearers for consultation at the mathematical reading-room of the university; those that are lithographed can be acquired by anybody, and I am much pleased to find them so well known here in America.

As another important point, I wish to say that I have always regarded my students not merely as hearers or pupils, but as collaborators. I want them to take an active part in my own researches; and they are therefore particularly welcome if they bring with them special knowledge and new ideas, whether these be original with them, or derived from some other source, from the teachings of other mathematicians. Such men will spend their time at Göttingen most profitably to themselves.

I have had the pleasure of seeing many Americans among my students, and gladly bear testimony to their great enthusiasm and energy. Indeed, I do not hesitate to say that, for some years, my higher lectures were mainly sustained by students whose home is in this country. But I deem it my duty to refer here to some difficulties that have occasionally arisen in connection with the coming of American students to Göttingen. Perhaps a frank statement on my part, at this opportunity, will contribute to remove these difficulties in part. What I wish to speak of is this. It frequently happens at Göttingen, and probably at other German universities as well, that American students desire to take the higher courses when their preparation is entirely inadequate for such work. A student having nothing but an elementary knowledge of the differential and integral calculus, usually coupled with hardly a moderate familiarity with the German language, makes a decided mistake in attempting to attend my advanced lectures. If he comes to Göttingen with such a preparation (or, rather, the lack of it), he may, of course, enter the more elementary courses offered at our university; but this is generally not the object of his coming. Would he not do better to spend first a year or two in one of the larger American universities? Here he would find more readily the transition to specialized studies, and might, at the same time, arrive at a clearer judgment of his own mathematical ability; this would save him from the severe disappointment that might result from his going to Germany.

I trust that these remarks will not be misunderstood. My presence here among you is proof enough of the value I attach to the coming of American students to Göttingen. It is in the interest of those wishing to go there that I speak; and for this reason I should be glad to have the widest publicity given to what I have said on this point.

Another difficulty lies in the fact that my higher lectures have frequently an encyclopedic character, conformably to the general tendency of my programme. This is not always just what is most needful to the American student, whose work is naturally directed to gaining the doctor's degree. He will need, in addition to what he may derive from my lectures, the concentration on a particular subject; and this he will often find best with other instructors, at Göttingen or elsewhere. I wish to state distinctly that I do not regard it as at all desirable that all students should confine their mathematical studies to my courses or even to Göttingen. On the contrary, it seems to me far preferable that the majority of the students should attach themselves to other mathematicians for certain special lines of work. My lectures may then serve to form the wider background on which these special studies are projected. It is in this way, I believe, that my lectures will prove of the greatest benefit.

In concluding I wish to thank you for your kind attention, and to give expression to the pleasure I have found in meeting here at Evanston, so near to Chicago, the great metropolis of this commonwealth, a number of enthusiastic devotees of my chosen science.

THE DEVELOPMENT OF MATHEMATICS AT THE GERMAN UNIVERSITIES.*

BY F. KLEIN.

THE eighteenth century laid the firm foundation for the development of mathematics in all directions. The universities as such, however, did not take a prominent part in this work; the *academies* must here be considered of prime importance. Nor can any fixed limits of nationality be recognized. At the beginning of the period there appears in Germany no less a man than *Leibniz;* then follow, among the kindred Swiss, the dynasty of the *Bernoullis* and the incomparable *Euler.* But the activity of these men, even in its outward manifestation, was not confined within narrow geographical bounds; to encompass it we must include the Netherlands, and in particular Russia, with Germany and Switzerland. On the other hand, under Frederick the Great, the most eminent French mathematicians, Lagrange, d'Alembert, Maupertuis, formed side by side with Euler and Lambert the glory of the Berlin Academy. The impulse toward a complete change in these conditions came from the French Revolution.

The influence of this great historical event on the development of science has manifested itself in two directions. On the one hand it has effected a wider separation of nations

* Translation, with a few slight modifications by the author, of the section *Mathematik* in the work *Die deutschen Universitäten,* Berlin, A. Asher & Co., 1893, prepared by Professor Lexis for the World's Columbian Exposition at Chicago.

with a distinct development of characteristic national qualities. Scientific ideas preserve, of course, their universality; indeed, international intercourse between scientific men has become particularly important for the progress of science; but the cultivation and development of scientific thought now progress on national bases. The other effect of the French Revolution is in the direction of educational methods. The decisive event is the foundation of the École polytechnique at Paris in 1794. That scientific research and active instruction can be directly combined, that lectures alone are not sufficient, and must be supplemented by direct personal intercourse between the lecturer and his students, that above all it is of prime importance to arouse the student's own activity,—these are the great principles that owe to this source their recognition and acceptance. The example of Paris has been the more effective in this direction as it became customary to publish in systematic form the lectures delivered at this institution; thus arose a series of admirable text-books which remain even now the foundation of mathematical study everywhere in Germany. Nevertheless, the principal idea kept in view by the founders of the Polytechnic School has never taken proper root in the German universities. This is the combination of the technical with the higher mathematical training. It is true that, primarily, this has been a distinct advantage for the unrestricted development of theoretical investigation. Our professors, finding themselves limited to a small number of students who, as future teachers and investigators, would naturally take great interest in matters of pure theory, were able to follow the bent of their individual predilections with much greater freedom than would have been possible otherwise.

But we anticipate our historical account. First of all we must characterize the position that Gauss holds in the science of this age. Gauss stands in the very front of the new development: first, by the time of his activity, his publications reaching back to the year 1799, and extending throughout the entire first half of the nineteenth century; then again, by the wealth of new ideas and discoveries that he has brought forward in almost every branch of pure and applied mathematics, and which still preserve their fruitfulness; finally, by his methods, for Gauss was the first to restore that *rigour* of demonstration which we admire in the ancients, and which had been forced unduly into the background by the exclusive interest of the preceding period in *new* developments. And yet I prefer to rank Gauss with the great investigators of the eighteenth century, with Euler, Lagrange, etc. He belongs to them by the universality of his work, in which no trace as yet appears of that specialization which has become the characteristic of our times. He belongs to them by his exclusively academic interest, by the absence of the modern teaching activity just characterized. We shall have a picture of the development of mathematics if we imagine a chain of lofty mountains as representative of the men of the eighteenth century, terminating in a mighty outlying summit,— *Gauss*,—and then a broader, hilly country of lower elevation; but teeming with new elements of life. More immediately connected with Gauss we find in the following period only the astronomers and geodesists under the dominating influence of *Bessel;* while in theoretical mathematics, as it begins henceforth to be independently cultivated in our universities, a new epoch begins with the second quarter of the present century, marked by the illustrious names of *Jacobi* and *Dirichlet.*

Jacobi came originally from Berlin and returned there for the closing years of his life (died 1851). But it is the period from 1826 to 1843, when he worked at Königsberg with *Bessel* and *Franz Neumann*, that must be regarded as the culmination of his activity. There he published in 1829 his *Fundamenta nova theoriæ functionum ellipticarum*, in which he gave, in analytic form, a systematic exposition of his own discoveries and those of Abel in this field. Then followed a prolonged residence in Paris, and finally that remarkable activity as a teacher, which still remains without a parallel in stimulating power as well as in direct results in the field of pure mathematics. An idea of this work can be derived from the lectures on dynamics, edited by Clebsch in 1866, and from the complete list of his Königsberg lectures as compiled by Kronecker in the seventh volume of the *Gesammelte Werke.* The new feature is that Jacobi lectured exclusively on those problems on which he was working himself, and made it his sole object to introduce his students into his own circle of ideas. With this end in view he founded, for instance, the first mathematical seminary. And so great was his enthusiasm that often he not only gave the most important new results of his researches in these lectures, but did not even take the time to publish them elsewhere.

Dirichlet worked first in Breslau, then for a long period (1831–1855) in Berlin, and finally for four years in Göttingen. Following Gauss, but at the same time in close connection with the contemporary French scholars, he chose mathematical physics and the theory of numbers as the central points of his scientific activity. It is to be noticed that his interest is directed less towards comprehensive developments than towards simplicity of conception and questions of principle; these are also the considerations on which he insists particularly in his lectures. These lectures are characterized by perfect lucidity and a certain refined objectivity; they are at the same time particularly accessible to the beginner and suggestive in a high degree to the more advanced reader. It may be sufficient to refer here to his lectures on the theory of numbers, edited by Dedekind; they still form the standard text-book on this subject.

With Gauss, Jacobi, Dirichlet, we have named the men who have determined the direction of the subsequent development.

We shall now continue our account in a different manner, arranging it according to the universities that have been most prominent from a mathematical standpoint. For henceforth, besides the special achievements of individual workers, the principle of co-operation, with its dependence on local conditions, comes to have more and more influence on the advancement of our science. Setting the upper limit of our account about the year 1870, we may name the universities of *Königsberg*, *Berlin*, *Göttingen*, and *Heidelberg*.

Of Jacobi's activity at Königsberg enough has already been said. It may now be added that even after his departure the university remained a centre of mathematical instruction. *Richelot* and *Hesse* knew how to maintain the high tradition of Jacobi, the former on the analytical, the latter on the geometrical side. At the same time *Franz Neumann's* lectures on mathematical physics began to attract more and more attention. A stately procession of mathematicians has come from Königsberg; there is scarcely a university in Germany to which Königsberg has not sent a professor.

Of Berlin, too, we have already anticipated something in our account. The years from 1845 to 1851, during which *Jacobi* and *Dirichlet* worked together, form the culminating period of the first Berlin school. Besides these men the most prominent figure is that of *Steiner* (connected with the university from 1835 to 1864), the founder of the German synthetic geometry. An altogether original character, he was a highly effective teacher, owing to the one-sidedness with which he developed his geometrical conceptions. — As an event of no mean importance, we must here record the foundation (in 1826) of *Crelle's Journal für reine und angewandte Mathematik*. This, for decades the only German mathematical periodical, contained in its pages the fundamental memoirs of nearly all the eminent representatives of the rapidly growing science in Germany.

Among foreign contributions the very first volumes presented Abel's pioneer researches. *Crelle* himself conducted this periodical for thirty years; then followed *Borchardt*, 1856–1880; now the Journal has reached its 110th volume. — We must also mention the formation (in 1844) of the *Berliner physikalische Gesellschaft.* Men like *Helmholtz*, *Kirchhoff*, and *Clausius* have grown up here; and while these men cannot be assigned to mathematics in the narrower sense, their work has been productive of important results for our science in various ways. During the same period, *Encke* exercised, as director of the Berlin astronomical observatory (1825–1862), a far-reaching influence by elaborating the methods of astronomical calculation on the lines first laid down by Gauss. — We leave Berlin at this point, reserving for the present the account of the more recent development of mathematics at this university.

The discussion of the *Göttingen school* will here find its appropriate place. The permanent foundation on which the mathematical importance of Göttingen rests is necessarily the Gauss tradition. This found, indeed, its direct continuation on the physical side when *Wilhelm Weber* returned from Leipsic to Göttingen (1849) and for the first time established systematic exercises in those methods of exact electro-magnetic measurement that owed their origin to Gauss and himself. On the mathematical side several eminent names follow in rapid succession. After Gauss's death, Dirichlet was called as his successor and transferred his great activity as a teacher to Göttingen, for only too brief a period (1855–59). By his side grew up *Riemann* (1854–66), to be followed later by *Clebsch* (1868–72).

Riemann takes root in Gauss and Dirichlet; on the other hand he fully assimilated Cauchy's ideas as to the use of complex variables. Thus arose his profound creations in the theory of functions which ever since have proved a rich and permanent source of the most suggestive material. Clebsch sustains, so to speak, a complementary relation to Riemann. Coming originally from Königsberg, and occupied with mathematical physics, he had found during the period of his work at Giessen (1863–68) the particular direction which he afterwards followed so successfully at Göttingen. Well acquainted with the work of Jacobi and with modern geometry, he introduced into these fields the results of the algebraic researches of the English mathematicians Cayley and Sylvester, and on the double foundation thus constructed, proceeded to build up new approaches to the problems of the entire theory of functions, and in particular to Riemann's own developments. But with this the significance of Clebsch for the development of our science is not completely characterized. A man of vivid imagination who readily entered into the ideas of others, he influenced his students far beyond the limits of direct instruction; of an active and enterprising character, he founded, together with C. Neumann in Leipsic, a new periodical, the *Mathematische Annalen,* which has since been regularly continued, and is just concluding its 41st volume.

We recall further those memorable years of Heidelberg, from 1855 to perhaps 1870. Here were delivered Hesse's elegant and widely read lectures on analytic geometry. Here Kirchhoff produced his lectures on mathematical physics. Here, above all, Helmholtz completed his great papers on mathematical physics, which in their turn served as basis for Kirchhoff's elegant later researches.

It remains now to speak of the *second Berlin school,* beginning also about the middle of the century, but still operating upon the present age. *Kummer, Kronecker, Weierstrass,* have been its leaders, the first two, as students of Dirichlet, pre-eminently engaged in developing the theory of numbers, while the last, leaning more on Jacobi and Cauchy, became, together with Riemann, the creator of the modern theory of functions. Kummer's lectures can here merely be named in passing; with their clear arrangement and exposition they have always proved especially useful to the majority of students, without being particularly notable for their specific contents. Quite different is the case of Kronecker and Weierstrass, whose lectures became in the course of time more and more the expression of their scientific individuality. To a certain extent both have thrust intuitional methods into the background and, on the other hand, have in a measure avoided the long formal developments of our science, applying themselves with so much the keener criticism to the fundamental analytical ideas. In this direction Kronecker has gone even farther than Weierstrass in trying to banish altogether the idea of the irrational number, and to reduce all developments to relations between integers alone. The tendencies thus characterized have exerted a wide-felt influence, and give a distinctive character to a large part of our present mathematical investigations.

We have thus sketched in general outlines the state reached by our science about the year 1870. It is impossible to carry our account beyond this date in a similar form. For the developments that now arise are not yet finished; the persons whom we should have to name are still in the midst of their creative activity. All we can do is to add a few remarks of a more general nature on the present aspect of mathematical science in Germany. Before doing this, however, we must supplement the preceding account in two directions.

Let it above all be emphasized that even within the limits here chosen, we have by no means exhausted the subject. It is, indeed, characteristic of the German universities that their life is not wholly centralized, — that wherever a leader appears, he will find a sphere of activity. We may name here, from an earlier period, the acute analyst *J. Fr. Pfaff*, who worked in Helmstädt and Halle from 1788 to 1825, and, at one time, had Gauss among his students. Pfaff was the first representative of the *combinatory* school, which, for a time, played a great rôle in different German universities, but was finally pushed aside in the manifold development of the advancing science. We must further mention the three great geometers, *Möbius* in Leipsic, *Plücker* in Bonn, *von Staudt* in Erlangen. Möbius was, at the same time, an astronomer, and conducted the Leipsic observatory from 1816 till 1868. Plücker, again, devoted only the first half of his productive period (1826–46) to mathematics, turning his attention later to experimental physics (where his researches are well known), and only returning to geometrical investigation towards the close of his life (1864–68). The accidental circumstance that each of these three men worked as teacher only in a narrow circle has kept the development of modern geometry unduly in the background in our sketch. Passing beyond university circles, we may be allowed to add the name of *Grassmann*, of Stettin, who, in his *Ausdehnungslehre* (1844 and 1862), conceived a system embracing the results of modern geometrical speculation, and, from a very different field, that of *Hansen*, of Gotha, the celebrated representative of theoretical astronomy.

We must also mention, in a few words, the *development of technical education*. About the middle of the century, it became the custom to call mathematicians of scientific eminence to the polytechnic schools. Foremost in this respect stands Zürich, which, in spite of the political boundaries, may here be counted as our own; indeed, quite a number of professors have taught in the Zürich polytechnic school who are to-day ornaments of the German universities. Thus the ideal of the Paris school, the combination of mathematical with technical education, became again more prominent. A considerable influence in this direction was exercised by *Redtenbacher's* lectures on the theory of machines which attracted to Carlsruhe an ever-increasing number of enthusiastic students. Descriptive geometry and kinematics were scientifically elaborated. *Culmann* of Zürich, in creating graphical statics, introduced the principles of modern geometry, in the happiest manner, into mechanics. In connection with the scientific advance thus outlined, numerous new polytechnic schools were founded in Germany about 1870 and during the following years, and some of the older schools were reorganized. At Munich and Dresden, in particular, in accordance with the example of Zürich, special departments for the training of teachers and professors were established. The polytechnic schools have thus attained great importance for mathematical education as well as for the advancement of the science. We must forbear to pursue more closely the many interesting questions that present themselves in this connection.

If we survey the entire field of development described above, this, at any rate, appears as the obvious conclusion, in Germany as elsewhere, that the number of those who have an earnest interest in mathematics has increased very rapidly and that, as a consequence, the amount of mathematical production has grown to enormous proportions. In this respect an imperative need was supplied when *Ohrtmann* and *Müller* established in Berlin (1869) an annual bibliographical review, *Die Fortschritte der Mathematik*, of which the 21st volume has just appeared.

In conclusion a few words should here be said concerning the modern development of university instruction. The principal effort has been to reduce the difficulty of mathematical study by improving the seminary arrangements and equipments. Not only have special seminary libraries been formed, but study rooms have been set aside in which these libraries are immediately accessible to the students. Collections of mathematical models and courses in drawing are calculated to disarm, in part at least, the hostility directed against the excessive abstractness of the university instruction. And while the students find everywhere inducements to specialized study, as is indeed necessary if our science is to flourish, yet the tendency has at the same time gained ground to emphasize more and more the mutual interdependence of the different special branches. Here the individual can accomplish but little; it seems necessary that many co-operate for the same purpose. Such considerations have led in recent years to the formation of a German mathematical association (*Deutsche Mathematiker-Vereinigung*). The first annual report just issued (which contains a detailed report on the development of the theory of invariants) and a comprehensive catalogue of mathematical models and apparatus published at the same time indicate the direction that is here to be followed. With the present means of publication and the continually increasing number of new memoirs, it has become almost impossible to survey comprehensively the different branches of mathematics. Hence it is the object of the association to collect, systematize, maintain communication, in order that the work and progress of the science may not be hampered by material difficulties. Progress itself, however, remains — in mathematics even more than in other sciences — always the right and the achievement of the individual.

GÖTTINGEN, January, 1893.

ISBN: 0-8218-2733-2

9 780821 827338

CHEL/339.H